Although the effects of exercise and mechanical forces on musculoskeletal and cardiovascular systems have been well documented, the actual mechanisms by which mechanical forces act at the cellular level are not well understood. At present, studies of the interaction of mechanical forces with cells encompass many different cell types in various tissues. This volume draws together these seemingly disparate observations and makes comparisons between the nature of the cellular responses in different tissues. Studies of cells derived from skeletal muscle, bone and cardiovascular tissue are considered, providing a comprehensive synthesis and review of recent work. The volume will be of interest to all those working in musculoskeletal and cardiovascular biology, as well as those taking courses in exercise and sports science, biomechanics and orthopaedics.

T0297160

SOCIETY FOR EXPERIMENTAL BIOLOGY
SEMINAR SERIES: 54

BIOMECHANICS AND CELLS

SOCIETY FOR EXPERIMENTAL BIOLOGY SEMINAR SERIES

A series of multi-author volumes developed from seminars held by the Society for Experimental Biology. Each volume serves not only as an introductory review of a specific topic, but also introduces the reader to experimental evidence to support the theories and principles discussed, and points the way to new research.

BIOMECHANICS AND CELLS

Edited by

F. Lyall
Department of Obstetrics and Gynaecology, University of Glasgow

A. J. El Haj
Department of Biology, University of Birmingham

CAMBRIDGE
UNIVERSITY PRESS

CAMBRIDGE UNIVERSITY PRESS
Cambridge, New York, Melbourne, Madrid, Cape Town, Singapore, São Paulo, Delhi

Cambridge University Press
The Edinburgh Building, Cambridge CB2 8RU, UK

Published in the United States of America by Cambridge University Press, New York

www.cambridge.org
Information on this title: www.cambridge.org/9780521114547

First published 1994
This digitally printed version 2009

A catalogue record for this publication is available from the British Library

Library of Congress Cataloguing in Publication data
Biomechanics and cells/edited by R. Lyall and A. J. El Haj.
    p. cm. – (Society for Experimental Biology Seminar series; 54)
    Includes bibliographical references and index.
    ISBN 0 521 45454 9
    1. Biomechanics. 2. Tissues–Mechanical properties. I. Lyall, F. II. El Haj,
Alicia. III. Series: Seminar series (Society for Experimental Biology (Great
Britain)); 54.
QH513.B555 1994
611'.018–dc29   93-42046   CIP

ISBN 978-0-521-45454-4 hardback
ISBN 978-0-521-11454-7 paperback

Contents

Contributors

ABBOTT, J.
Electro-Biology Inc., 6 Upper Pond Road, Parsippany, New Jersey 07054-1079, USA
BEE, J. A.
Royal Veterinary College, Royal College Street, London NW1 0TU, UK
BINDERMAN, I.
Sackler Faculty of Medicine, Tel Aviv University, 6 Weizman Street, Tel Aviv 64239 Israel
BINGMANN, D.
Institut für Physiologie, Universitäts Gesamthochschule Essen, Germany
BLACKSHAW, S. E.
Department of Cell Biology, University of Glasgow, Glasgow G12 8QQ, UK
BODIN, P.
Department of Anatomy and Developmental Biology, University College London, Gower Street, London WC1E 6BT, UK
BURGER, E. H.
Department of Oral Cell Biology, ACTA/Vrije University, Van der Boechorststraat 7, 1081 BT Amsterdam, The Netherlands
BURNSTOCK, G.
Department of Anatomy and Developmental Biology, University College London, Gower Street, London WC1E 6BT, UK
BUTTERWORTH, P.
The University of Surrey, Guildford, Surrey, UK
CLARKE, N.
Department of Veterinary Basic Sciences, The Royal Veterinary College, Royal College Street, London NW1 0TU, UK
COHEN, C. R.
Department of Surgery, Yale University School of Medicine, 333 Cedar Street, New Haven, Connecticut 06510, USA

CURTIS, A. S. G.
Department of Cell Biology, University of Glasgow, Glasgow G12 8QQ, UK
DEEHAN, M. R.
MRC Blood Pressure Unit and Department of Medicine and Therapeutics, Western Infirmary, Glasgow G11 6NT, UK
DU, W.
Department of Surgery, Yale University School of Medicine, 333 Cedar Street, New Haven, Connecticut 06510, USA
EL HAJ, A. J.
School of Biological Sciences, The University of Birmingham, P.O. Box 363, Birmingham B15 2TT, UK
EVANS, L.
Department of Surgery, Yale University School of Medicine, 333 Cedar Street, New Haven, Connecticut 06510, USA
FERMOR, B. F.
Department of Anatomy, School of Veterinary Science, University of Bristol, Park Row, Bristol BS1 5LS, UK
GERLACH, G.-F.
Department of Anatomy and Developmental Biology, Royal Free Hospital School of Medicine, University of London, Rowland Hill Street, London NW3 2PF, UK
GOLDSPINK, G.
Department of Anatomy and Developmental Biology, Royal Free Hospital School of Medicine, University of London, Rowland Hill Street, London NW3 2PF, UK
GRIFFITH, T. M.
Department of Diagnostic Radiology, University of Wales College of Medicine, Heath Park, Cardiff CF4 4XN, UK
HESLOP-HARRISON, J. S.
Karyobiology Group, John Innes Centre for Plant Science Research, Colney Lane, Norwich NR4 7UJ, UK
HUTCHESON, I. R.
Department of Diagnostic Radiology, University of Wales College of Medicine, Heath Park, Cardiff CF4 4XN, UK
ISALES, C.
Department of Surgery, Yale University School of Medicine, 333 Cedar Street, New Haven, Connecticut 06510, USA
JAENICKE, T.
Department of Anatomy and Developmental Biology, Royal Free Hospital School of Medicine, University of London, Rowland Hill Street, London NW3 2PF, UK

JONES, D. B.
Cell Biology Laboratory, Experimental Orthopaedics, Westfalische Wilhelms Universität, Domagkstrasse 3, D-4400 Münster, Germany
KLEIN-NULEND, J.
Department of Oral Cell Biology, ACTA/Vrije University, Van der Boechorststraat 7, 1081 BT Amsterdam, The Netherlands
LANYON, L. E.
The Royal Veterinary College, University of London, Royal College Street, London NW1 0TU, UK
LEIVSETH, G.
Akademie für Manuelle Medizin, Universität Münster, Domagkstrasse 3, D-48129 Münster, Germany
LIU, H.-X.
Department of Veterinary Basic Sciences, The Royal Veterinary College, Royal College Street, London NW1 0TU, UK
LOESCH, A.
Department of Anatomy and Developmental Biology, University College London, Gower Street, London WC1E 6BT, UK
LYALL, F.
Department of Obstetrics and Gynaecology, Queen Elizabeth Building, Royal Infirmary, Glasgow G31 2ER, UK
McCULLOCH, C. A. G.
Faculty of Dentistry, University of Toronto, Ontario, Canada
MILLS, I.
Department of Surgery, Yale University School of Medicine, 333 Cedar Street, New Haven, Connecticut 06510, USA
MILNER, P.
Department of Anatomy and Developmental Biology, University College London, Gower Street, London WC1E 6BT, UK
PENDER, N.
Clinical Orthodontics, Department of Clinical Dental Sciences, The University of Liverpool, Liverpool L69 3BX, UK
REIPERT, B.
CRC Department of Physics and Instrumentation, Paterson Institute for Cancer Research, Christie Hospital, Wilmslow Road, Manchester M20 9BX, UK
ROSALES, O. R.
Department of Surgery, Yale University School of Medicine, 333 Cedar Street, New Haven, Connecticut 06510, USA
SAWADA, Y.
Laboratory for Cell Biology, Experimental Orthopaedics, Westfalische Wilhelms Universität, Domagkstrasse 3, D-4400 Münster, Germany

SKERRY, T. M.
Department of Anatomy, School of Veterinary Science, University of Bristol, Park Row, Bristol BS1 5LS, UK
SUMPIO, B. E.
Department of Surgery, Yale University School of Medicine, 333 Cedar Street, New Haven, Connecticut 06510, USA
THOMAS, G. P.
Department of Biology, University of Birmingham, P.O. Box 363, Birmingham B15 2TT, UK
VAN DER SLOTEN, J.
Department of Biomechanica, Catholic University of Leuven, Heverlee, Belgium
VELDHUIJZEN, J. P.
Department of Oral Cell Biology, ACTA/Vrije University, Van der Boechorststraat 7, 1081 BT Amsterdam, The Netherlands

Part 1
Soft tissue

B. E. SUMPIO, W. DU, C. R. COHEN,
L. EVANS, C. ISALES, O. R. ROSALES
and I. MILLS

Signal transduction pathways in vascular cells exposed to cyclic strain

The importance of external physical forces in influencing the biology of cells is just being realised. Recent reports demonstrate that exposure of endothelial cells (EC) to a flowing culture media or to repetitive elongation can result in changes in morphology, proliferation and secretion of macromolecules (Dewey *et al.*, 1981; Davies *et al.*, 1984; Frangos, Eskin & McIntire, 1985; Sumpio *et al.*, 1987; Diamond, Eskin & McIntire, 1989; Sumpio & Widmann, 1990; Iba & Sumpio, 1991, 1992). Now, the major impetus in the field is to define the 'mechanosensor(s)' on the cells that are sensitive to the different external forces, the coupling intracellular pathways and the subsequent nuclear events which precede the cell response.

Mechanosensors

Cell surface sensors

The cell's plasma membrane, besides serving as a barrier to protect the cell interior, is the site of action and translation of external to internal signals. Although no 'strain-receptor' as such has been identified, it is clear that endothelial cells can 'sense' changes in pressure and strain. Furthermore, it is likely that this 'sensor' is located on the cell surface. The endothelial cell surface consists of multiple projections covered by a thin layer of glycocalyx (consisting mainly of glycoproteins, proteoglycans and derived substances). In an effort to characterise possible cell surface sensors, Suarez & Rubio (1991) perfused isolated guinea pig hearts with concavalin A or heparinase (agents which modify endothelial cell surface glycoproteins) and attenuated both the flow and pressure stretch induced rise in glycolytic flux normally seen in guinea pig hearts, while having no effect on basal glycolytic values. In contrast,

Society for Experimental Biology Seminar Series 54: *Biomechanics and Cells*, ed. F. Lyall & A. J. El Haj. © Cambridge University Press 1994, pp. 3–22.

4 B. E. SUMPIO *et al.*

infusion with hyaluronidase and chondroitinase (agents which modify surface proteoglycans) had no effect on flow-activated glycolysis, while modifying basal glycolytic values. Their data suggest that cell surface glycoproteins might play a role in 'sensing' changes in blood vessel flow and pressure.

Ion channels

Ion channels maintain the electrochemical balance, pH and osmolarity of the interior milieu of the cell. These channels span the cell membrane and are therefore subject to the mechanical stresses which affect the membrane. It would seem reasonable to presume that any force which affects the membrane tension may affect the channels within.

One model of mechanotransduction involving ion channels pictures the channel as a 'cylindrical plug' of protein embedded in the membrane (Kirber, Walsh & Singer, 1988) and predicts that in order for energy to activate the channel, it needs to be transmitted by cytoskeletal strings. Experiments utilising cytochalasins to disrupt the cytoskeleton have demonstrated an increase in channel sensitivity. The cytochalasins are postulated to act by cleaving *either* non-channel attachment sites in the membrane, thereby increasing the lattice spacing, or the parallel elastic elements in the membrane. The cell shape of capillary endothelial cells has also been shown to be determined by a counterbalance between the contractile forces of the microfilaments and the compression resistance of the microtubules (Ingber & Folkman, 1989). It suggests that membrane tension, and hence the state of the mechanosensitive ion channels, may be a function of the cytoskeletal elements.

Recent studies have also suggested the presence of mechanosensitive ion channels in vascular cells (for review see Davies, 1989). Endothelial cells have been shown to have 'stretch-activated' calcium-permeable channels using the cell attached patch technique. Using porcine aortic endothelial cells Lansman, Hallam & Rink (1987) found that by applying suction (10–20 mm Hg or 1.3–2.7 dynes cm^{-2}) to a cell attached membrane patch, a cation selective current (with a slope conductance of 19.1 pS and reversal potential of 17 mV when the patch electrode contained 110 mM $CaCl_2$) was observed which they speculated could carry sufficient calcium to serve as the second messenger in prostacyclin and EDRF (endothelium-derived relaxing factor) release. Stretch-activated ion channels have been described in other cell types (Guharay & Sachs, 1984; Morris, 1990; Davis *et al.*, 1992), including rat ventricular myocytes (Craelius, Chen & El-Sherif, 1988) and arterial baroreceptors (Schreiber *et al.*, 1971; Singh, 1982). The demonstration of a cation-selective channel that is permeable to Ca^{2+} is not a universal finding since others have failed to confirm these channels in cultured endothelial cells from either bovine pulmonary artery, human umbilical vein or rabbit aorta (Adams *et al.*, 1989). In addition, Morris & Horn (1991) have recently suggested that, at least in the snail neurone growth cones, the single channel recordings of a potassium-selective stretch-activated channel may be an irrelevant artefact. Since they were unable to correlate their single-channel recordings with any macroscopic currents, it is as yet unclear whether these findings will pertain exclusively to snail neurones. In contrast, Davis *et al.* (1992) have reported that they were able to observe activation of a cation channel ($K^+ > Na^+ > Ba^{2+} > Ca^{2+}$; with a slope conductance of 7 pS when there was 110 mM $CaCl_2$ in the patch electrode) in both whole cell and single-channel recordings when 10–15% stretch (above control) was applied to porcine coronary artery smooth muscle cells.

Adragna (1991) studied the effect of cyclic stretch on sodium and potassium transport in bovine aortic EC (BAEC). He found an increase in Na^+ and K^+ content in cells stretched at 3 cycles min^{-1} and 24% strain for 7 days. The effect of ouabain and of ouabain plus furosemide suggested that cyclic stretch stimulated entry, or inhibited exit of Na^+ and K^+, or both. Further experiments using bradykinin suggested that cyclic stretch and bradykinin act on endothelial cells via an increase in intracellular calcium through release from intracellular pools and calcium entry into the cell. The intracellular calcium then behaves as a second messenger activating various ion channels to produce the observed ion shifts.

Further studies are required to investigate the importance and effect of mechanosensitive ion channels in vascular cells, which are subject to a continuously changing mechanical environment.

Intracellular coupling pathways

Once the endothelial cell has 'sensed' a change in the applied strain, the signal is 'transduced' to the cell interior. Multiple signal transduction pathways exist in vascular endothelial cells. This is not unexpected since the endothelium is metabolically active and many different inputs are processed simultaneously. However, it would be unlikely that one could activate a single second messenger pathway in an isolated manner, without this having important effects on the other pathways. For example, a rise in intracellular calcium can activate the calcium/calmodulin-dependent enzyme constitutive nitric oxide synthase, which in turn will stimulate the soluble guanylate cyclase and increase cGMP levels; this in turn may

Cellular Calcium Homeostasis in the Vascular Endothelial Cell

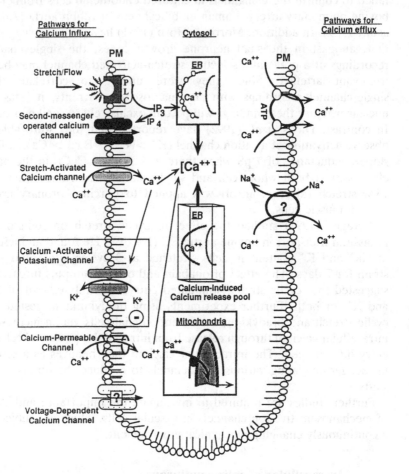

Fig. 1. Cytosolic calcium concentration is the result of a balance between calcium influx, intracellular calcium redistribution and calcium efflux. There are multiple pathways for calcium influx in endothelial cells, some or all of which may or may not be present depending on the type of endothelial cell studied. Possible pathways for influx include: (i) stretch or flow activated increases in inositol phosphate. IP$_4$ may then open a calcium permeable channel; (ii) there may be a stretch-activated calcium channel; (iii) a rise in [Ca^{2+}]$_i$ through mobilisation of intracellular stores may then activate a calcium-dependent potassium channel. By hyperpolarising the EC it favours calcium entry through a calcium-permeable channel, down its electrogenic gradient; (iv) voltage activated calcium channel, though present in many excitable cells does

feedback and modulate intracellular calcium and cAMP levels via the cGMP-dependent cAMP phosphodiesterase (Lincoln, 1989; Lugnier & Schini, 1990; Lewis & Smith, 1992; Schilling & Elliot, 1992). Thus, cellular responses could be viewed as the result of the integrated activation of a variety of signal transduction pathways; abnormalities in cellular response would then be viewed as a persistent or abnormal activation of one of these pathways, ultimately leading to disease (atherosclerosis, hypertension, etc.).

Intracellular calcium

In many tissues an elevation in intracellular calcium is an important second messenger system. However, since a persistent elevation of intracellular calcium is potentially toxic to the cell, a rise in cytosolic calcium (whether from intracellular or extracellular sources) is balanced by an increase in calcium efflux (Fig. 1). Since the cell membrane has a low calcium permeability, calcium must enter through specific pathways. In general, pathways for calcium entry include the membrane calcium-permeable channels which are receptor or second messenger operated, or voltage-dependent.

Initial reports (Singer & Peach, 1982; Rubanyi, Schwartz & Vanhoutte, 1985) suggested that the endothelium contained dihydropyridine-sensitive channels, implicating voltage dependent calcium channels as signal transducers in this tissue. However, multiple subsequent studies (Bossu *et al.*, 1989; Takeda & Klepper, 1990; Busse, Lückhoff & Pohl, 1992; Schilling & Elliott, 1992) using both whole cell or single channel patch-clamp studies have been unable to demonstrate that voltage dependent calcium channels are present in the vascular endothelium. The reason(s) for these discrepancies are unclear but may relate to differences in cell preparation. Interestingly, recent reports in renal tubule cells (Bacskai & Friedman, 1990; Gesek & Friedman, 1992) suggest the possibility that one can induce the appearance of latent dihydropyridine-sensitive calcium channels by pre-exposure to certain

---

Fig. 1 (*cont.*)

not appear to be present in EC. There seem to be at least two intracellular calcium pools an IP_3 mobilisable pool and a calcium-inducible pool. The efflux pathways for calcium efflux are not well characterised but calcium can be pumped out through a calcium ATPase pump or sodium–calcium exchanger. The latter has been shown to be involved in calcium influx in certain cells, although this does not seem to be the case for EC.

hormones, which may account for some of the different results reported in endothelial cells. While cell membrane potential plays a role in calcium entry, in endothelial cells it is membrane hyperpolarisation and not depolarisation that stimulates calcium influx. A rise in cytosolic calcium through mobilisation of intracellular stores activates a calcium-dependent potassium channel, which by hyperpolarising the cell favours calcium entry down its electrogenic gradient (Busse *et al.*, 1992). It has been reported that arterially derived endothelial cells have a more negative resting potential (-60 to -70 mV) while venous endothelial cells are more depolarised (-30 mV) and lack the inward rectifying K^+ current (Jacob, Sage & Rink, 1990). These differences, together with species differences, may partly account for some of the reported differences in calcium influx reported from the various endothelial cell types.

The pathways for calcium entry during cyclic strain include a 'stretch-activated' cation channel. In addition, we have observed changes in intracellular calcium in bovine aortic EC exposed to cyclic strain (O. Rosales, unpublished observations). Application of 24% strain at 60 cycles min^{-1} leads to a rapid rise in intracellular calcium through an increase in IP_3 (inositol 1,4,5-trisphosphate) with mobilisation of internal calcium stores and through stimulating extracellular calcium influx. The calcium influx is blocked by gadolinium and removal of extracellular calcium, suggesting entry through a cation channel. Endothelial cells have also been recently reported to have an inositol 1,3,4,5-tetrakisphosphate (IP_4) modulated low conductance channel (Lückhoff & Clapham, 1992). Its contribution to the changes we have observed is unclear although IP_4 must be generated during cyclic strain.

Taken together these data suggest that in tissues subjected to cyclic strain (i.e. the blood vessel) calcium mobilisation from intracellular stores and calcium entry through stretch-activated channels may play an important second messenger role.

Inositol phosphates and diacylglycerols

The plasma membrane contains a pool of lipid precursor molecules which upon hormonal activation can serve as substrates for the generation of second messengers (Liscovitch, 1992). For instance, the phosphatidyl-inositols serve as precursors for a number of different messenger molecules. Stimulation of cell-surface receptors initiates hydrolysis of phosphatidylinositol 4,5-bisphosphate (PIP_2), which produces at least two second messengers – IP_3 and diacylglycerol (DAG).

The effect of cyclic strain on the activation of inositol phosphates and DAG in bovine aortic EC has been a subject of active investigation in our laboratory. Rosales & Sumpio (1992a) recently reported that the initiation of cyclic strain from 0 to 60 cycles min^{-1} (0.5 s elongation alternating with 0.5 s relaxation) induced a time-dependent, monophasic accumulation of IP$_3$ (peak at 10 s), inositol bisphosphate (IP$_2$), and inositol monophosphate (IP$_1$) on EC prelabelled with [^{3}H]myoinositol. Cyclic strain had a similar effect on radioimmunoactive IP$_3$ mass.

Changes in cyclic strain frequency promoted inositol phosphate hydrolysis as well. When bovine aortic EC were subjected to an acute variance in cyclic strain frequency from 60 to 100 cycles min^{-1}, a sequential increase in IP$_3$, IP$_2$ and IP$_1$ levels paralleled those triggered by the initiation of cyclic strain. An increase in IP$_3$ mass was documented at the same time. The kinetics of DAG formation nearly paralleled those of IP$_3$ production in both types of experiments, suggesting that the early changes in IP$_3$ and DAG levels were the direct result of PIP$_2$ hydrolysis mediated by phospholipase C. The increase in intracellular calcium from the IP$_3$-sensitive pool and higher levels of DAG provide for some of the substrates needed for PKC (protein kinase C) activation.

Although the effect of cyclic strain on the kinetics of IP$_3$ levels in BAEC was characterised by a single early-phase peak at 10 s and a rapid return to baseline levels after 35 s, the generation of DAG was biphasic. There was an early peak at 10 s followed by a sustained phase of activation after 100 s that persisted for up to 8 min. In a variety of cell types, DAG formation is biphasic with an initial peak as a result of PIP$_2$ hydrolysis by a phosphatidylinositol-specific phospholipase C (PI-PLC). This DAG is normally transient, and temporally corresponds to the formation of IP$_3$ which is frequently followed by a more sustained elevation of DAG. This later sustained phase of DAG formation is most likely due to hydrolysis of phosphatidylcholine (PC) in various stimulated cells.

The sustained formation of DAG in BAEC in response to cyclic strain is crucial because sustained PKC activation is a prerequisite essential for causing long-term physiological response such as cell proliferation and differentiation. Several mechanisms may be responsible for the signal-induced formation of DAG from PC (Griendling *et al.*, 1986; Liscovitch, 1992). There is a PC-reactive PLC which requires tyrosine phosphorylation for its activation. This type of PLC remains poorly defined. It is more likely that PC is hydrolised by phospholipase D (PLD) in a signal-dependent manner resulting in the formation of phosphatidic acid, which is converted to DAG by the removal of its phosphate. Phospholipase A$_2$ (PLA$_2$) has been shown to be activated by most of the signals which induce PI hydrolysis. Thus, the sustained phase accumulation emphasises the multiple sources of DAG in EC, including

PIP$_2$, and ensures that those cellular responses that are controlled by PKC can be maintained throughout agonist stimulation. Although phospholipases A$_2$, C and D are all present in the endothelial cell and all of them appear to be activated by cyclic strain, the initial rapid rise of IP$_3$ (the IP$_3$ peak is observed within 10 s of stretch) is secondary to PI-PLC activation (Evans, unpublished observations). The role of each of these signal-activated phospholipases in the generation of DAG is under current investigation in our laboratory.

The underlying mechanisms that modulate this monophasic increase in IP$_3$ levels are yet to be determined. We have preliminary evidence that the transient rise in IP$_3$ formation is not modified by calcium-free media, nickel, an inhibitor of calcium influx, or the benzohydroquinone tBu-BHQ, a specific mobiliser of the IP$_3$ sensitive pool (O. Rosales, unpublished observations). Similarly, pretreatment with charybdotoxin, a non-selective K$^+$ channel blocker, or gadolinium, a stretch-activated calcium channel antagonist, failed to reduce the IP$_3$ response to cyclic strain. It is not blocked by pertussis toxin (1 μg ml^{-1} per 4 h) which would suggest that this response is not G$_i$-linked. One possibility, since G$_q$ is linked to PI-PLC in other tissues, is that stretch is activating PI-PLC through this same G protein. These observations suggest that cyclic strain-mediated PIP$_2$ hydrolysis is controlled by phospholipase C, yet it is extracellular calcium and pertussis toxin independent. Further work, however, needs to be done in this area.

Protein kinase C (PKC)

PKC is a family of serine/threonine kinases characterised by a dependence on phospholipids and DAG for activation. Since their original description (Kawahara *et al.*, 1980; Kishimoto *et al.*, 1980) these kinases have been implicated in numerous cellular responses (see Rasmussen *et al.*, 1991 for review). The present dogma is that PKC appears to be predominantly in a cytosolic fraction and cell activation is associated with translocation of PKC to a membrane fraction where it is then in an active form (although to date no study has demonstrated that PKC translocation is synonymous with its activation and this sequence of events has been questioned). Part of the present confusion is a result of the multiple methods used to study its function: (i) one can examine its cellular redistribution into either cytosolic or membrane fractions (e.g. by Western blotting, where the fractions are separated by centrifugation, separated by SDS-PAGE and immunoblotted with specific antibodies); (ii) PKC translocation from cytosol to membrane can be evaluated by immunocytochemistry where using electron or confocal microscopy with isoform specific PKC antibodies one can determine exactly where PKC is located; (iii) activity can be measured by phosphorylation of a PKC-specific substrate: the 80 kDa myristoylated, alanine-rich, C-kinase substrate or MARCKS (Albert *et al.*, 1986; Calle *et al.*, 1992); (iv) one can indirectly assess PKC's role in a particular response by use of 'specific' inhibitors. The usefulness of the last approach is seriously hampered, however, by the lack of selective PKC inhibitors (Rasmussen *et al.*, 1991).

At least nine isoforms of PKC have been identified with the α, β (1 and 2) and γ isoforms, being dependent on calcium for activation (Bell & Burns, 1991). What role an increase in cytosolic calcium plays in modulating PKC is still unclear but calcium probably facilitates the initial translocation of PKC to the membrane and, once translocated, continued calcium influx also appears to modulate the activity of the membrane-bound fraction. In contrast to the α, β and γ isoforms, the ε, δ and ξ forms appear to be calcium-independent. Both groups of kinases have two identifiable domains: a catalytic domain (the ATP binding site, blockable by staurosporine) and a regulatory domain (the phospholipid and DAG binding site, blockable by calphostin) (Bell & Burns, 1991; Rasmussen *et al.*, 1991).

Available data suggest that PKC is an important mediator of the adaptation of vascular EC to cyclic strain *in vitro* (Rosales & Sumpio, 1992b). PKC activity was documented by measuring the transfer of phosphate from [^{32}P]ATP to histone. Cyclic deformation (60 cycles min^{-1}, 24% maximum strain) resulted in a biphasic translocation of PKC from the cytosolic to the particulate fraction. There was an early increase in activity in the particulate fraction at 10 s which paralleled those of IP$_3$ and DAG, and a second rise which occurred after 100 s of cyclic strain and was sustained up to nearly 8 min. The activation of PKC is pivotal in the modulation of multiple EC responses to mechanical stretch, such as the secretion of prostacyclin, endothelin EDRF, tissue plasminogen activator, growth and proliferation. This relocation in PKC activity was confirmed by immunohistochemical staining and confocal microscopy. In resting BAEC, PKCα and β predominantly stained the cytosol (in contrast to HUVECS (human umbilical vein endothelial cells) where by Western blotting the α, ε and ζ isoforms are present). Cyclic strain caused an intense perinuclear and nuclear translocation of PKCβ and a marked perinuclear halo redistribution of PKCα (Fig. 2). In contrast, stimulation of BAEC with phorbol ester resulted in only PKCα redistribution to the perinuclear region, whereas PKCβ staining remained mostly cytosolic

Basal   Stretch

PKC α

PKC β

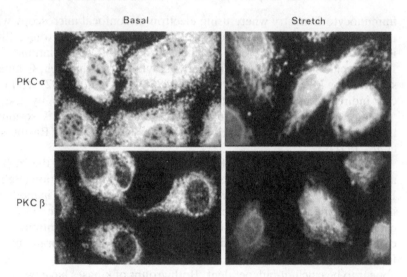

Fig. 2. Confocal fluorescence microscopy of protein kinase C localisation in BAEC. BAEC were exposed to 0.1% DMSO for 10 min, fixed,
and incubated with antibodies against PKC α, β for 2 h, and subsequently exposed to the rhodamine-labelled anti-mouse IgG for 60 min.
Images were focused at the middle portion of the nucleus. Bovine aortic
endothelial cells contain two PKC isoforms, α and β. In response to
cyclic strain both of these isoforms could be translocated (Rosales &
Sumpio, 1992b).

(Rosales *et al.*, 1992) (Fig. 3). The selective relocation of the different
PKC isoenzymes in the adaptation of vascular cells to haemodynamic
forces is currently a subject of active investigation.

PKC also seems to regulate the growth of BAEC in the presence and
absence of cyclic strain. The mitogenic activity of mechanical stress
(Sumpio *et al.*, 1987; Rosales & Sumpio, 1992b) was abrogated by the
PKC inhibitor calphostin C. Preincubation with 0.05 μM calphostin C
(PKC_{IC-50}) resulted in a decreased rate of proliferation of BAEC subjected to cyclic strain or stationary culture conditions for 5 days. These
observations suggest that PKC is probably a major signal transducer of
the effects of cyclic strain on the vascular endothelium. Further studies
aimed to identify the intracellular targets of cyclic strain-mediated PKC
activation will be fundamental to enhance our understanding of how
mechanical stress modifies the cell cycle.

Basal   100 nM TPA

PKC α

PKC β

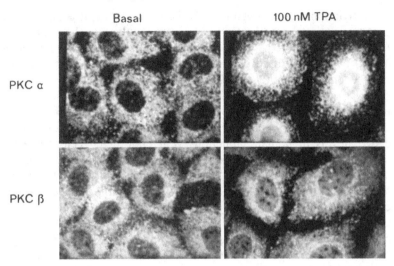

Fig. 3. Exposure of BAEC to 100 nM TPA in 0.1% DMSO for 10 min, resulted in a shift of PKC α to the perinuclear region, whereas the fluorescent pattern of PKC β remained predominantly cytosolic (Rosales *et al.*, 1992).

Adenylate cyclase

Strain-induced activation of adenylate cyclase has been demonstrated in many cell types using very different strain models (Schreiber *et al.*, 1971; Somjen *et al.*, 1980; Singh, 1982; Watson, Haneda & Morgan, 1989). In our laboratory, we have shown that basal and forskolin stimulated adenylate cyclase activities were stimulated by acute cyclic strain in bovine aortic endothelial cells (Letsou *et al.*, 1990). Adenylate cyclase activity increased progressively from 1 to 5 min and returned to basal by 7 min. Interestingly, this increase in adenylate cyclase activity was associated with a transient loss of $G_{1\alpha 1,2}$ immunoreactivity over a similar time course (Cohen, Mills & Sumpio, 1992; Mills, Cohen & Sumpio, 1992). This phenomenon was specific for EC $G_{1\alpha 1,2}$ since $G_{s\alpha}$ immunoreactivity was unchanged as were $G_{1\alpha 1,2}$ and $G_{s\alpha}$ in bovine aortic smooth muscle cells. The loss of $G_{1\alpha 1,2}$ immunoreactivity was dependent on probing with antisera directed at the carboxy terminus of $G_{1\alpha 1,2}$. Due to the long half-life of the G protein subunits (Silbert *et al.*, 1990), it is unlikely that a change in the G protein α subunit degradation rate is involved. Since the carboxy terminus of $G_{1\alpha 1,2}$ has a CAAX motif (C = cysteine, A = aliphatic amino acid, X = terminal amino acid) that is a common site of methylation, isoprenylation and ribosylation (Clarke *et al.*, 1988), it is possible that a post-translational modification of $G_{1\alpha1,2}$ is involved in the strain-induced loss of immunoreactivity. It remains to be determined whether acute cyclic strain of cultured bovine aortic endothelial cells is a stimulus for these events or whether novel pathways are activated.

It was our hypothesis that intracellular cAMP levels would be elevated in endothelial cells subject to cyclic strain, since adenylate cyclase was stimulated. Surprisingly, we have been unable to detect an increase in cAMP levels in human saphenous vein endothelial cells subjected to acute cyclic strain (Iba, Mills & Sumpio, 1992). These experiments were performed in the presence of 5 mM IBMX (3-isobutyl-1-methylxanthine) to prevent the degradation of cAMP. Furthermore, in order to rule out compartmentalisation and/or release of cAMP, we measured levels in a crude membrane fraction and the medium and were unable to measure changes in cAMP levels in response to cyclic strain. In a perfused guinea pig heart model exposed to an acute haemodynamic overload, Schreiber *et al.* (1971) also demonstrated an increase in adenylate cyclase activity that was not accompanied by an elevation in intracellular cAMP levels.

The failure to detect strain-induced elevation in cAMP accumulation may be based on our choice of studying this response in endothelial cells derived from saphenous vein. In recent studies performed in our laboratory, we observed that bovine aortic EC subjected to cyclic strain exhibit a stimulation of cAMP-dependent protein kinase activity (PKA) (Cohen, Mills & Sumpio, 1993). Thus for this response as for cell proliferation and tPA secretion, it appears that the source of endothelial cells under study will dictate the observed response (Iba *et al.*, 1990).

The effect of cyclic strain on adenylate cyclase activity has been examined extensively in coronary vascular smooth muscle cells. In contrast to that described above for EC, coronary vascular smooth muscle cells exhibit a reduction in adenylate cyclase activity in response to acute cyclic strain (Mills *et al.*, 1990). Both basal and maximal stimulation in the presence of forskolin and Mn^{2+} were reduced by approximately 30% in membranes obtained in stretched versus unstretched cells. Interestingly, basal adenylate cyclase activity was inversely related to the degree of cyclic strain. A pressure of -20 kilopascals was required to demonstrate a statistically significant effect although a trend in the reduction of adenylate cyclase activity was observed at -15 kPa.

Similar findings were obtained in coronary vascular smooth muscle cells exposed to chronic stretch (Wiersbitsky, Mills & Gerwitz, 1991). Basal, Gpp(NH)p and forskolin stimulated adenylate cyclase activity were all inhibited significantly in stretched (1 day) versus unstretched cells. The reduction in adenylate cyclase activity observed after 1 day of stretch was associated with a significant reduction in the levels of $G_{s\alpha45}$, the alpha subunit of the stimulatory G protein. In contrast, the alpha subunit of the inhibitory G protein, $G_{1\alpha1,2}$, remained unchanged after 1 day of cyclic strain. With continued cyclic strain of 7 days, a significant reduction in $G_{s\alpha45}$ levels remained, compared with unstretched cells. However, strain-induced inhibition of adenylate cyclase activity was not observed at this later time point. Thus, while strain-induced changes in $G_{s\alpha45}$ levels may play an important role in temporarily modulating adenylate cyclase activity, over time other factors are effective in restoring adenylate cyclase activity to control levels.

Nuclear events

One possible hypothesis that links the phosphatidylinositol/PKC and adenylate cyclase/cAMP/PKA pathways to subsequent nuclear events is that the activation of PKC or PKA is followed by phosphorylation of target proteins such as transcription factors which could potentially change DNA binding affinities and transcriptional activities (Bohman, 1990). These transcription factors may in turn trigger a transient increase in the activity of early response genes such as the nuclear oncogenes c-*fos* and c-*myc*, which would then encode specific transcription factors that may induce the expression of final target genes. Although this hypothesis is consistent with previous observations by other groups, i.e. c-*fos* regulating the α-actin gene, further studies will be necessary to clarify the role of PKC and PKA in strain-mediated events in EC.

Proto-oncogenes

Sadoshima *et al.* (1992), examining cardiac cells plated in a silicone substrate, found that a 20% static stretch induced hypertrophy of these cells. In contrast, they found that exposure of non-myocyte cells (including endothelial cells) to stretch under the same conditions led to proliferation with a significant increase in cell number. In addition the rapid response genes (c-*fos*, c-*jun*, c-*myc*, JE and Egr-1) are induced by stretch (20%) in the myocytes. This initial activation is followed by induction of 'fetal' genes: skeletal α-actin, β-myosin heavy chain and atrial natriuretic factor. These investigators have characterised the 'stretch response element' for c-*fos* as being located within 356 base pairs of the 5' flanking region and that of the fetal genes as being 3412 and 628 bp outside of the 5' flanking region. Using vascular smooth muscle cells, Bhalla & Sharma (1993) have also demonstrated that when these cells are exposed to cyclical strain at 60 cycles min^{-1}, these cells demonstrated an increase in [^{3}H]thymidine incorporation and induction of c-*fos* (with a peak at 30 min of stretching and a decline to baseline by 2 h) and elastin genes in a PKC-dependent manner. We have previously reported that cyclical strain is associated with an increase in proliferation of BAEC. Recent work in our laboratory also has found that cyclical strain is associated with induction of the rapid response elements c-*fos* and c-*jun* in BAEC (Du, Xu & Sumpio, 1993). These data suggest that mechanical strain is a specific activator of gene expression which is likely to play a role in vascular remodelling.

Transcription factors

Of current intense research interest is the identification and characterisation of transcriptional regulators which cause the observed changes in gene expression in EC exposed to external forces. Cross-talk between signalling pathways is known to occur (Berridge, 1987; Gilman, 1989; Karin, 1992). By comparing activation of specific genes with cyclic strain or with other chemical stimuli, such as phorbol ester, or even with other external forces such as shear stress, a better understanding of the specificity and selectivity of stimuli–nuclear coupling can be obtained. Genes respond to diverse stimuli through specific regions within their promoters. These regions contain a motif to which cellular proteins termed transcription factors bind and enhance the transcription level of the genes. Repressor binding sites also exist, which function to shut down gene transcription. A given chemical stimulus can generate specific transcription factors that may bind to the same promoter element in different genes. Thus, shear stress or cyclic strain may also ultimately induce gene expression in EC through the interaction of specific transcriptional factors with specific responsive elements in cyclic strain-responsive genes.

Dimerization of the family of *fos* and *jun* proteins takes place through a structural motif termed the 'leucine zipper'. *fos-jun* or *jun-jun* dimers form transcriptional activators which can bind AP-1 promoter sites [5'-TGAGTCAG-3'] of various genes. Activation of the cAMP/PKA pathway leads to binding of cAMP response element-binding protein to the CRE promoter sites. Preliminary evidence from our laboratory (Du *et al.*, 1993) and from others (Resnik *et al.*, 1993) indicates that the cyclic strain or shear stress stimuli can be coupled to EC response via activation of PKC, PKA and c-*fos* and c-*jun*, with the formation of transcriptional activators which regulate EC genes containing AP-1 and CRE promoter sites.

The study of transcriptional regulators would help correlate the large data base of sequenced genes with patterns of protein expression in EC, such as endothelin and IL-6. Furthermore, the relationship of specific responsive elements to nuclear elements which might be activated upon exposure of EC to chemical or mechanical forces could be analysed.

Summary

Figure 4 provides a simplified scheme of signal transduction pathways in endothelial cells that may be involved in coupling cyclic strain to gene expression. The mechanosensor(s) is still unknown. It is likely that intracellular calcium levels are increased by direct activation of calcium channels and by IP_3-mediated calcium release. Protein kinase C and protein kinase A have been shown to be activated but the role of the Ca^{2+}/calmodulin kinases still needs to be verified. Transcription factors AP-1, NF-kB and CREB are generated and can result in activation of specific gene transcripts.

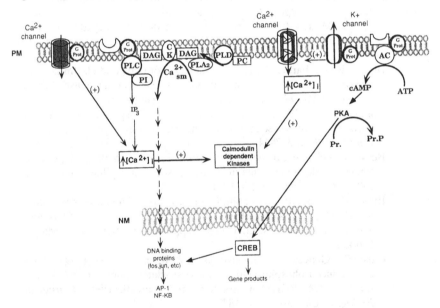

Fig. 4. A simplified scheme of signal transduction pathways in endothelial cells that may be involved in coupling cyclic strain to gene expression.

References

Adams, D., Barakeh, J., Laskey, R. & Breemen, C. V. (1989). Ion channels and regulation of intracellular calcium in vascular endothelial cells. *FASEB Journal* 3, 2389–400.

Adragna, N. C. (1991). Cyclic stretching affects sodium and potassium transport in vascular endothelial cell and its response to bradykinin. *Journal of Vascular Medicine and Biology* 3, 60–7.

Albert, K., Walaas, S., Wang, J.-T. & Greengard, P. (1986). Widespread occurrence of '87 kDa', a major specific substrate for protein kinase C. *Proceedings of the National Academy of Sciences USA* 83, 2822–6.

Bacskai, B. J. & Friedman, P. A. (1990). Activation of latent Ca^{2+} channels in renal epithelial cells by parathyroid hormone. *Nature* 347, 388–91.

Bell, R. & Burns, D. (1991). Lipid activation of protein kinase C. *Journal of Biological Chemistry* 266, 4661–4.

Berridge, M. J. (1987). Inositol trisphosphate and diacylglycerol: two interacting second messengers. *Annual Review of Biochemistry* 56, 159–93.

Bhalla, R. & Sharma, R. (1993). Induction of c-fos and elastin gene in response to mechanical stretch of vascular smooth muscle cells. *Journal of Vascular Medicine and Biology* (in press).

Bohmann, D. (1990). Transcription factor phosphorylation: a link between signal transduction and the regulation of gene expression. *Cancer Cells* 2, 337–44.

Bossu, J., Feltz, A., Rodeau, J. & Tanzi, F. (1989). Voltage-dependent transient calcium currents in freshly dissociated capillary endothelial cells. *FEBS Letters* 255, 377–80.

Busse, R., Lückhoff, A. & Pohl, U. (1992). Role of membrane potential in EDRF synthesis and release. In *Endothelial Regulation of Vascular Tone*, ed. U. Ryan & G. Rubanyi, pp. 105–20. New York: Marcel Dekker.

Calle, R., Ganesan, S., Smallwood, J. & Rasmussen, H. (1992). Glucose-induced phosphorylation of myristoylated alanine-rich C kinase substrate (MARCKS) in isolated rate pancreatic islets. *Journal of Biological Chemistry* 267, 18723–7.

Clarke, S., Vogel, J., Deschenes, R. & Stock, J. (1988). Post-translational modification of the Ha-ras oncogene protein: Evidence for a third class of protein carboxyl methyltransferases. *Proceedings of the National Academy of Sciences USA* 85, 4643–7.

Cohen, C. R., Mills, I. & Sumpio, B. E. (1992). Are G-proteins mechanotransducers for endothelial cells? *Surgical Forum* 43, 327–9.

Cohen, C. R., Mills, I. & Sumpio, B. E. (1993). The activity of cyclic

AMP dependent protein kinase (PKA) is increased by acute cyclic stretch in endothelial cells *in vitro*. *FASEB Journal* **7**, 786A.

Craelius, W., Chen, V. & El-Sherif, N. (1988). Stretch activated ion channels in ventricular myocytes. *Bioscience Reports* **8**, 407–14.

Davies, P., Forbes-Dewey, J. C., Bussolari, S., Gordon, E. & Gimbrone, M. A. (1984). Influence of hemodynamic forces on vascular endothelial function. *Journal of Clinical Investigation* **73**, 1121–9.

Davies, P. F. (1989). How do vascular endothelial cells respond to flow? *News in Physiological Sciences* **4**, 22–5.

Davis, M., Donovitz, J. & Hood, J. (1992). Stretch-activated single-channel and whole cell currents in vascular smooth muscle cells. *American Journal of Physiology* **262**, C1083–8.

Dewey, C. F., Bussolari, S. R., Gimbrone, M. A. & Davies, P. F. (1981). The dynamic response of vascular endothelial cells to fluid shear stress. *Journal of Biochemistry (England)* **103**, 177–85.

Diamond, S., Eskin, S. & McIntire, L. (1989). Fluid flow stimulates tissue plasminogen activator secretion by cultured human endothelial cells. *Science* **243**, 1483–5.

Du, W., Xu, W. & Sumpio, B. E. (1993). Proto-oncogenes c-*fos*, c-*jun* and transcription factors AP-1 and NF-kB are activated in endothelial cells exposed to cyclic strain. *FASEB Journal* **7**, 2A.

Frangos, J., Eskin, S. & McIntire, L. (1985). Flow effects on prostacyclin production by cultured human endothelial cells. *Science* **227**, 1477–9.

Gesek, F. & Friedman, P. (1992). On the mechanisms of parathyroid hormone stimulation of calcium uptake by mouse distal convoluted tubules cells. *Journal of Clinical Investigation* **90**, 749–58.

Gilman, A. G. (1989). G proteins and regulation of adenylyl cyclase. *Journal of the American Medical Association* **262**, 1819–25.

Griendling, K. K., Rittenhouse, S. E., Brock, T. A., Ekstein, L. S., Gimbrone, J. M. A. & Alexander, R. W. (1986). Sustained diacylglycerol formation from inositol phospholipids in angiotensin II-stimulated vascular smooth muscle cells. *Journal of Biological Chemistry* **261**, 5901–6.

Guharay, F. & Sachs, F. (1984). Stretch-activated single ion channel currents in tissue cultured embryonic chick skeletal muscle. *Journal of Physiology* **352**, 685–701.

Iba, T., Maitz, S., Furbert, T. & Sumpio, B. E. (1990). Effect of cyclic stretch on endothelial cells from different vascular beds. *Circulatory Shock* **35**, 193–8.

Iba, T., Mills, I. & Sumpio, B. (1992). Intracellular cyclic AMP levels in endothelial cells subjected to cyclic strain *in vitro*. *Journal of Surgical Research* **52**, 625–30.

Iba, T. & Sumpio, B. E. (1991). Morphologic evaluation of human endothelial cells subjected to repetitive stretch *in vitro*. *Microvascular Research* **42**, 245–54.

Iba, T. & Sumpio, B. (1992). Tissue plasminogen activator expression in endothelial cells exposed to cyclic strain. *Cell Transplantation* **1**, 43–50.

Ingber, D. & Folkman, J. (1989). Tension and compression as basic determinants of cell form and function: utilization of a cellular tensegrity mechanism. In *Cell Shape: Determinants, Regulation and Regulatory Role*, ed. W. D. Stein & F. Bronner, pp. 3–31. San Diego: Academic Press.

Jacob, R. Sage, S. & Rink, T. (1990). Aspects of calcium signalling in endothelium. In *The Endothelium: An Introduction to Current Research*, ed. J. Warren, pp. 33–44. New York: Wiley-Liss.

Karin, M. (1992). Signal transduction from cell surface to nucleus in development and disease. *FASEB Journal* **6**, 2581–90.

Kawahara, Y., Takai, Y., Minakuchi, R., Sano, K. & Nishizuka, Y. (1980). Possible involvement of Ca^{2+} activated, phospholipid dependent protein kinase in platelet activation. *Journal of Biochemistry* **88**, 913–16.

Kirber, M., Walsh, J. & Singer, J. (1988). Stretch-activated ion channels in smooth muscle a mechanism for initiation of stretch-induced contraction. *Pflügers Archiv* **412**, 339–46.

Kishimoto, A., Takai, Y., Mori, T., Kikkawa, U. & Nishizuka, Y. (1980). Activation of calcium and phospholipid-dependent protein kinase by diacylglycerol, its possible relation to phosphatidylinositol turnover. *Journal of Biological Chemistry* **255**, 2273–6.

Lansman, J., Hallam, T. & Rink, T. (1987). Single stretch-activated ion channels in vascular endothelial cells as mechanotransducers? *Nature* **325**, 811–13.

Letsou, G., Rosales, O., Maitz, S., Vogt, A. & Sumpio, B. (1990). Stimulation of adenylate cyclase activity in cultured endothelial cells subjected to cyclic stretch. *Journal of Cardiovascular Surgery* **31**, 634–9.

Lewis, M. & Smith, J. (1992). Factors regulating the release of endothelium-derived relaxing factor. In *Endothelial Regulation of Vascular Tone*, ed. U. Ryan & G. Rubayani, pp. 139–54. New York: Marcel Dekker.

Lincoln, T. (1989). Cyclic GMP and mechanisms of vasodilation. *Pharmacological Therapy* **41**, 479–502.

Liscovitch, M. (1992). Cross-talk among multiple signal-activated phospholipases. *Trends in Biological Sciences*, 393–9.

Lückhoff, A. & Clapham, D. (1992). Inositol 1,3,4,5-tetrakisphosphate activates an endothelial Ca^{2+}-permeable channel. *Nature* **355**, 356–8.

Lugnier, C. & Schini, V. (1990). Characterization of cyclic nucleotide phosphodiesterases from cultured bovine aortic endothelial cells. *Biochemical Pharmacology* **39**, 75N-<84.

Mills, I., Cohen, C. & Sumpio, B. (1992). Modulation of $G_{ia1,2}$ immuno-reactivity in endothelial cells subjected to cyclic strain. *FASEB Journal* **6**, 1355A.

Mills, I., Letsou, G., Rabban, J., Sumpio, B. & Gerwtz, J. (1990). Mechanosensitive adenylate cyclase activity in coronary artery vascular smooth muscle. *Biochemical Biophysical Research Communications* **171**, 143–7.

Morris, C. (1990). Mechanosensitive ion channels. *Journal of Membrane Biology* **113**, 93–107.

Morris, C. & Horn, R. (1991) Failure to elicit neuronal macroscopic mechanosensitive currents anticipated by single-channel studies. *Science* **251**, 1246–9.

Rasmussen, H., Calle, R., Ganesan, S., Smallwood, J., Throckmorton, D. & Zawalich, W. (1991). Protein kinase C: role in sustained cellular responses. In *Protein Kinase C: Current Concepts and Future Perspectives*, ed. R. Epand & D. Lester. Chichester, England: Ellis Horwood.

Resnick, N., Collins, T., Atkinson, W., Bonthron, D. T., Dewey, C. F., Jr & Gimbrone, M. A., Jr (1993). Platelet-derived growth factor B chain promoter contains a *cis*-acting fluid shear-stress-responsive element. *Proceedings of the National Academy of Sciences USA* **90**, 4591–5.

Rosales, O. & Sumpio, B. (1992a). Changes in cyclic strain increase inositol trisphosphate and diacylglycerol in endothelial cells. *American Journal of Physiology* **262**, C956–62.

Rosales, O. & Sumpio, B. (1992b). Protein kinase C is a mediator of the adaptation of vascular endothelial cells to cyclic strain *in vitro*. *Surgery* **112**, 459–66.

Rosales, O. R., Isales, C., Nathanson, M. & Sumpio, B. E. (1992). Immunocytochemical expression and localization of protein kinase C in bovine aortic endothelial cells. *Biochemical and Biophysical Research Communications* **189**, 40–6.

Rubanyi, G., Schwartz, A. & Vanhoutte, P. (1985). The calcium agonists BAY K 8644 and (+)202,791 stimulate the release of endothelial relaxing factor from canine femoral arteries. *European Journal of Pharmacology* **117**, 143–4.

Sadoshima, J., Jahn, L., Takahashi, T., Kulik, T. & Izumo, S. (1992). Molecular characterization of the stretch-induced adaptation of cultured cardiac cells. *Journal of Biological Chemistry* **267**, 10551–60.

Schilling, W. & Elliott, S. (1992). Ca^{2+} signalling mechanisms of vascular endothelial cells and their role in oxidant-induced endothelial cell dysfunction. *American Journal of Physiology* **262**, H1617–30.

Schreiber, S., Klein, I., Oratz, M. & Rothschild, M. (1971). Adenylate cyclase activity and cyclic AMP in acute cardiac overload: a method of measuring cyclic AMP production based on ATP specific activity. *Journal of Molecular and Cellular Cardiology* **2**, 55–65.

Silbert, S., Michel, T., Lee, R. & Neer E. (1990). Differential degradation rates of the G protein α in cultured cardiac and pituitary cells. *Journal of Biological Chemistry* **265**, 3102–4.

Singer, H. & Peach, M. (1982). Calcium- and endothelium mediated vascular smooth muscle relaxation in rabbit aorta. *Hypertension* **4**, II19–25.

Singh, J. (1982). Stretch stimulates cyclic nucleotide metabolism in the isolated frog ventricle. *Pflügers Archiv* **395**, 162–4.

Somjen, D., Binderman, I., Berger, E. & Harell, A. (1980). Bone remodelling induced by physical stress is prostaglandin E_2 mediated. *Biochimica et Biophysica Acta* **627**, 91–100.

Suarez, J. & Rubio, R. (1991). Regulation of glycolytic flux by coronary flow in guinea pig heart. Role of vascular endothelial cell glycocalyx. *American Journal of Physiology* **261** H1994–2000.

Sumpio, B., Banes, A., Levin, L. & Johnson, G. (1987). Mechanical stress stimulates aortic endothelial cells to proliferate. *Journal of Vascular Surgery* **6**, 252–6.

Sumpio, B. E., Banes, A. J., Buckley, M. & Johnson, G. (1988). Alterations in aortic endothelial cell morphology and cytoskeletal protein synthesis during cyclic tensional deformation. *Journal of Vascular Surgery* **7**, 130–8.

Sumpio, B. & Widmann, M. (1990). Enhanced production of an endothelium derived contracting factor by endothelial cells subjected to pulsatile stretch. *Surgery* **108**, 277–82.

Takeda, K. & Klepper, M. (1990). Voltage-dependent and agonist-activated ionic currents in vascular endothelial cells: a review. *Blood Vessels* **27**, 169–83.

Watson, P., Haneda, T. & Morgan, H. (1989). Effect of higher aortic pressure on ribosome formation and cAMP content in rat heart. *American Journal of Physiology* **256**, C1257–61.

Wiersbitsky, M., Mills, I. & Gerwitz, H. (1991). Stretch-induced reduction in G protein steady state levels in coronary vascular smooth muscle cells. *FASEB Journal* **5**, 1394A.

F. LYALL and M. R. DEEHAN

Effects of pressure overload on vascular smooth muscle cells

Hypertension, or high blood pressure, is an abnormal increase in the systemic arterial circulation usually caused by narrowing of the arterioles, the small-resistance vessels. The lumen may be reduced by structural changes of the vessel such as hypertrophy (cell enlargement) or hyperplasia (increased cell number), or by vasoconstriction caused by hormones and sympathetic nerve activity (Folkow, 1978; Lever, 1986). In this disorder the heart pumps a normal output of blood into a high-resistance circuit and blood pressure rises. Hypertension is associated with an increased risk of ischaemic heart disease and cerebrovascular disease. Hypertension is therefore an important disorder.

Mammalian arteries consist of an inner continuous monolayer of endothelium that is non-thrombogenic and serves as a structural semi-permeable barrier and intermediary between the blood and the underlying cells of the vessel wall. Smooth muscle cells (smc) lie within the media of the vessel and are responsible for the structural integrity of the vessel as well as vasomotor tone. Endothelial cells and smc exist in a quiescent state; however, in response to various stimuli they are capable of exuberant proliferation. As an example, when an artery is denuded of its endothelium, platelets adhere, thrombus formation is initiated, and both endothelial cells and smc are stimulated to proliferate (Ross, 1986). Endothelial proliferation is an effort to promote healing of the endothelium while smc proliferation occurs in response to mitogens in the vascular lumen. Smc reside in a dynamic state *in vivo*, being constantly subjected to pulsatile hydrostatic pressures and shear stresses; however, much of our knowledge of the biology of these cells has come from studies *in vitro*.

The cardiovascular system adapts to elevated blood pressure and thus haemodynamic load by remodelling (Folkow, 1975, 1978). Hypertension

Society for Experimental Biology Seminar Series 54: *Biomechanics and Cells*, ed. F. Lyall & A. J. El Haj. © Cambridge University Press 1994, pp. 23–36.

is also well recognised as a major risk factor for development of atherosclerotic disease (Freis, 1969). In hypertensive man the presence of left
ventricular hypertrophy considerably increases cardiac morbidity and
mortality. Imposition of increased load on the heart of mature animals
results in hypertrophy due to enlargement of myocytes and hyperplasia of
non-muscle components of the myocardium (Zak, 1974). Several studies
suggest that the association between hypertension and atherosclerosis
relates to increased vascular smooth muscle cell (vsmc) growth which is
a common feature of both diseases (Furuyama, 1962; Wolinsky, 1970;
Owens et al., 1981; Ross, 1981; Schwartz & Ross, 1984).

There are conflicting reports regarding the type of vsmc growth
response in experimental models of hypertension where smc are subjected to pressure overload. In slow-developing models of hypertension
such as the spontaneously hypertensive rat and the two-kidney, one-clip
Goldblatt model (Olivetti et al., 1982; Owens & Schwartz, 1982, 1983)
the major responses of vsmc, in large arteries, are hypertrophy and
hyperploidy (DNA replication) rather than hyperplasia. In a variety of
animal species aortic stenosis-induced hypertension elicits a rapid growth
response of vsmc in thoracic aortae but reports vary regarding the role
of vsmc hypertrophy versus hyperplasia. Vsmc DNA replication was
studied in rabbits made hypertensive by partial constriction of the abdominal aorta just proximal to the superior mesenteric artery (Bevan, Marthens & Bevan, 1976). In this study increased DNA synthesis and content
was reported in vessels proximal to the coarctation, which was interpreted as vsmc hyperplasia. However, Olivetti et al. (1980) found no
increase in smc number in thoracic aortas of rats following subdiaphrag-
matic constriction of the abdominal aorta and concluded that media
thickness was due to smc hypertrophy, not hyperplasia. The growth
response of vsmc in rats was examined following acute hypertension produced by partial ligation of the abdominal aorta between the renal arteries (Owens & Reidy, 1985). The frequency of vsmc undergoing DNA
replication was increased 25-fold in thoracic aortas while no differences
were observed in cells in abdominal aortic segments. By far the major
growth response was hyperplasia, with a 25% increase in medial vsmc
number. The marked contrast between results in this study and previous
studies by this group showing aortic medial hypertrophy in the spontaneously hypertensive and Goldblatt hypertensive rats was due to cellular
hypertrophy and hyperploidy without hyperplasia demonstrates that the
growth response of vsmc within a given blood vessel can be quite different depending on the model of hypertension.

The changes in conducting arteries are not seen prior to elevation
of blood pressure (Lee, 1985). Furthermore, antihypertensive treatment

(Limas, Westrum & Limas, 1980) prevents the structural modifications of the vessels, suggesting that these changes are secondary to hypertension. In contrast, structural alterations in the small muscular arteries and arterioles are present at the prehypertensive and developing hypertensive phases (Lee, 1985). There is indirect evidence that the rate of synthesis may be due to cellular factors that are genetically determined (Kanbe *et al.*, 1983; Lever, 1986; Mulvany, 1986).

To determine whether the increased thickness seen in media of mesenteric resistance vessels of Wistar–Kyoto rats made hypertensive by the Goldblatt procedure (one-kidney, one-clip model) was due to hypertrophy or hyperplasia of vsmc, the cellular dimensions of the vessels were measured using an unbiased stereological method. Furthermore, to investigate whether the changes seen could be secondary to the increased blood pressure, morphometric measurements were also made in renal arcuate arteries which, due to the constricting silver clip, had probably not been exposed to the increased pressure load (Korsgaard & Mulvany, 1988). The number of cell layers in the vessel wall was significantly increased in the mesenteric vessels although the number of cells was unaltered. No change was seen in the renal vessels. Thus these findings pointed towards hypertrophy not hyperplasia, with the increase in the number of cell layers due to the cells being packed around a smaller lumen. These findings contrasted with previous findings in resistance vessels in the spontaneously hypertensive rat (Lee *et al.*, 1983; Mulvany, Baandrup & Gundersen, 1985) and the genetically hypertensive Dahl strain rat (Lee & Triggle, 1986), where it was found that medial hypertrophy was caused by hyperplasia. In contrast to these results, the increased smooth muscle cell mass seen in the aortas of the spontaneously hypertensive rat (Owens *et al.*, 1981; Owens & Schwartz, 1982), renal hypertensive rat (Owens & Schwartz, 1983) and other experimental models of hypertension (Olivetti *et al.*, 1980) was found to be due to hypertrophy not hyperplasia. Thus it appears that hyperplasia in the resistance vessels of genetic models of hypertension could be associated with the primary factors causing the hypertension (Lee & Smeda, 1985), whereas changes in the aorta could be a response of smooth muscle cells to increased vascular wall stress (Wolinsky, 1970). The observation that renal arcuate arteries, due to the constricting clip, probably had not been exposed to increased pressure load and in these vessels no significant morphological changes were found, could be explained in two ways. First, there is increasing evidence for a trophic action of sympathetic nerves (Bevan, 1984) on vascular structure, therefore the differences may be due to differences in adrenergic activity between renal and mesenteric vessels in the renal hypertensive rat. Alternatively, the changes in the mesenteric vessels could be a direct response to the increased blood pressure and load on the cells. More recently, the effect of cyclic pressure loading was studied on blood vessels in the spontaneously hypertensive rat (Christensen, 1991). The results indicated that a reduction in pulse pressure and heart rate during antihypertensive treatment may be important in preventing the development of abnormal small artery structure in hypertension.

The cellular mechanisms that initiate and control adaptive cardiac and vascular changes *in vivo* are uncertain. The intracellular messenger systems implicated in cell growth include hydrolysis of inositol lipids and expression of proto-oncogenes (see the chapter by Sumpio *et al.*). *In vivo* studies have examined some of the early cellular signals that may be at work during the development of pressure-induced structural changes in the cardiovascular system when coarctation of the aorta is produced.

One study examined the effects of pressure overload on phosphoinositide hydrolysis and proto-oncogene expression in cardiovascular tissues (MacIver *et al.*, 1993). The hypothesis was that increased inositol phosphate metabolism and proto-oncogene expression are required for a growth response to increased pressure. The hypothesis was tested in the model of coarctation hypertension using four different cardiovascular tissues, the aorta proximal to the stenosis and mesenteric resistance vessels (all exposed to high blood pressure and thus increased load), the aorta below the coarctation and therefore exposed to low blood pressure and the left ventricle of the heart. Products of inositol lipid hydrolysis and levels of c-*myc*, c-*fos* and c-*ras* mRNAs were measured in vascular tissues after 72 h and 9 days after the induction of aortic coarctation in order to examine inositol phosphate and proto-oncogene signals during the development of pressure-related vascular structural changes. There was a significant increase in proximal aortic mass at both time points but no change in mesenteric resistance artery morphology in rats with coarctation. At 72 h there was a significant increase in c-*myc*, c-*fos* and H-*ras* mRNA which was accompanied by increased levels of (1,4,5)-trisphosphate in the proximal but not the distal aorta. In resistance arteries inositol phosphate production and proto-oncogene mRNA expression were unchanged. These results indicated that at 72 h coarctation induced structural changes in the proximal aorta and was associated with increased inositol phosphate production and stimulation of specific proto-oncogene mRNAs. Nine days following surgery, most of the structural changes in these tissues were completed and the raised signals were no longer observed. The results suggest that both inositol lipid hydrolysis and a rise in the expression of these proto-oncogenes are important processes in the development of vascular hypertrophy in this model of hypertension. The mesenteric resistance vasculature failed to show increased inositol production nor stimulation of proto-oncogenes at either time point. Furthermore, no structural change was seen during the experiment in the resistance arteries, despite the fact that this vascular bed rises above the coarctation and so was presumably exposed to a higher blood pressure and thus increased load. However, previous studies by this group have shown that morphological changes in this tissue only reach statistical significance 28 days after surgery (Izzard *et al.*, 1991). The authors concluded that stimulation of these second messenger pathways may mediate vascular refashioning and suggested that alternative mechanisms may be involved in mesenteric vessel thickening where slower structural development is observed.

It is clear from the literature described above that studying the direct effects of mechanical forces *in vivo* is complex since it is difficult to know whether changes observed in smc growth and second messengers result as a direct consequence of increased blood pressure and thus haemodynamic load on the tissue under investigation, or whether the effects are due to involvement of humoral factors.

To determine whether these early signalling events, i.e. proto-oncogene expression and phosphoinositide turnover are increased as a direct result of increased pressure, we have developed an *in vitro* system capable of subjecting vsmc from mesenteric arteries in culture to a fixed linear stretch. The advantage of this cell culture system is that mechanical effects on second messenger systems can be determined independently of humoral factors. The system is composed of a perspex base containing four vertical screws onto which 0.5 mm silicone sheets (11 × 10 cm²) can be mounted. Two perspex blocks are placed at each end of the silicone and are further secured by tightening the screws. The silicone adopts a rectangular well shape in which vsmc are grown. A perspex lid is placed directly over the well to maintain sterile conditions. In preparation for experiments the perspex pieces of the cell stretching apparatus were sterilised by soaking in 70% ethanol overnight and then air dried in a laminar flow hood. The silicone and thus the vsmc attached to the silicone could be stretched by up to 20% by turning the handle as illustrated in Fig. 1. Silicone sheets were coated with fibronectin for optimal vsmc growth.

To show that the vsmc, attached to the fibronectin coated silicone (FCS), were stretched by the same amount as the silicone, vsmc were plated sparsely onto the fibronectin treated silicone and a single unstretched cell was photographed. The silicone membrane was then stretched by 20% and the cell was photographed again. The length of the unstretched cell was 17.5 μm. After stretching the silicone by 20% the length of the cell increased to 21 μm. This corresponded to a 20%

Fig. 1. Cell stretching apparatus.

increase in the length of the cell. Thus this was conclusive proof that the cell was being stretched by the same amount as the silicone.

In preparation for experiments vsmc were plated at a density of 0.8×10^6 cells onto fibronectin coated silicone, grown to confluence in 10% fetal calf serum/10% horse serum in Dulbecco's modified Eagle medium, and then made quiescent by incubation in serum-free medium for 24 h. The medium was then replaced with 10 ml of fresh serum-free medium, the silicone stretched by 20%, and the cells incubated for a predetermined time at 37 °C. After the appropriate period of stretch, the medium was removed from the cells and cold serum-free medium was added. The cells were scraped from the silicone and pelleted by centrifugation. Total inositol phosphates and proto-oncogene c-*fos* mRNA levels were measured as described previously (Lyall *et al.*, 1992). Figure 2 is a representative autoradiograph showing the effects of a 20% stretch on c-*fos* mRNA expression in confluent quiescent vsmc. Cells were stretched for 0, 15, 30, 60, 180 and 360 min. There was no c-*fos* mRNA detectable in unstimuiated cells; however, in response to a 20% stretch there was a rapid accumulation of c-*fos* mRNA, which was maximal at 15 min, still detectable at 30 min, and then declined to basal levels. The effects of a 20% stretch on total inositol phosphates released from vsmc is shown in

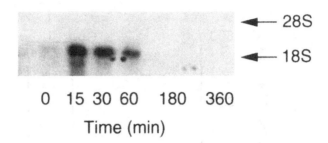

0 15 30 60   180   360

Time (min)

Fig. 2. A representative autoradiograph showing the effects of a fixed stretch of 20% on c-*fos* mRNA induction in vsmc. Vsmc were plated at a density of 0.8×10^6 cells and grown to confluence. Cells were serum-deprived for 24 h and then stretched by 20% for 0, 15, 30, 60, 180 or 360 min. RNA was extracted from the cells, run on agarose gels and blotted onto nylon membranes as described in the text. The membranes were hybridised with a v-*fos* cDNA probe.

Fig. 3. 10% FCS and 10^{-7} M angiotensin II were included as positive controls. Results are expressed as a percentage of control $\pm$ SE where control = 100%. Incubation with 10% FCS resulted in a 7.7-fold increase in inositol phosphate levels released. Incubation with 10^{-7} M angiotensin II resulted in a 2.3-fold increase in inositol phosphates released. After stretching cells by 20% a 3.2-fold increase in inositol phosphate levels was obtained.

The time course of c-*fos* mRNA expression reached maximal levels by 15 min, when the cells were stretched by 20%. This time course of c-*fos* mRNA expression is similar to that observed when growth factors such as angiotensin II are added to cultured vsmc (Naftilan *et al.*, 1989; Taubman *et al.*, 1989; Lyall *et al.*, 1992). 10% FCS and 10^{-7} M angiotensin II resulted in a 7.7- and 2.3-fold increase, respectively in inositol phosphate levels released. Stretching the vsmc by 20% resulted in a 3.2-fold increase in inositol phosphate levels released when compared to unstretched vsmc. Therefore stretch and growth factors induce common cell signalling pathways, events which are thought to be important in the initiation of vsmc growth. In summary, our data demonstrate that mechanical stretch increases both proto-oncogene expression and phosphoinositide turnover in vsmc *in vitro*, without participation of humoral factors. Thus it is possible that mechanical stretch may initiate vascular hypertrophy and/or hyperplasia via these signalling pathways.

Smooth muscle cells are the predominant cell type in the media of blood vessel walls and they produce a variety of components that com-

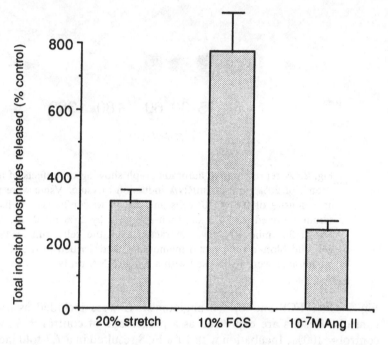

Fig. 3. Effect of 20% stretch, 10% FCS and 10^{-7} M angiotensin II on phosphoinositide turnover in vsmc. The cells were plated at a density of 0.8×10^6 cells on fibronectin coated silicone membranes, grown to confluence, serum deprived for 24 h, labelled with [³H]inositol and then either stretched by 20% or stimulated with 10% FCS or 10^{-7} M angiotensin II. Total [³H]inositol phosphates released were determined as described in the text. Results are expressed as a percentage of control ± SE where control = 100%. $n = 6$.

prise the extracellular matrix (Burke & Ross, 1979). This connective tissue matrix forms the foundation of the vessel wall and plays an important role in its mechanical properties. Qualitative and quantitative relationships have been described between the composition of arterial walls and estimates of medial stress. These findings suggest that physical forces relating to pressure and flow direct medial cell biosynthesis, thereby modulating structural adaptations to haemodynamic changes. The mechanisms regulating tissue adaptations to physical stresses are poorly understood. The production of matrix molecules by vsmc has been shown to be quantitatively and qualitatively altered during hypertension and atherosclerosis (Wolinsky, 1970; Mayne 1984). Collagen is one of the primary components of the smc matrix and has been shown to be altered when tensile stress has been applied to the vessel wall. Vsmc maintained in culture produce many of the components of the extracellular matrix including collagen. The effects of physical forces on collagen synthesis by cultured vsmc was investigated in response to a 3 cycles min^{-1} force regime (Sumpio *et al.*, 1988a). The stress unit used in this system consisted of a vacuum unit connected to a regulator solenoid valve that is controlled by a computer with a timer programme. The timer controlled the duration and frequency of the applied stress or relaxation. Cells were cultured on flexible-bottomed culture plates with a hydrophilic surface that could be deformed by vacuum. The degree of deformation was controlled by regulating the vacuum level. Since the cells were attached to the surface of the culture dish, they were assumed to be subjected to the same force as the plate bottom. For these experiments the smc in culture were subjected to cycles of 10 s of a maximum 25% elongation and 10 s of relaxation for 5 days. In this system pulsatile stretch was shown to increase collagen and non-collagen protein synthesis. Leung, Glagov & Mathews (1977) subjected smc grown on elastin membranes to repeated elongation and relaxation (52 cycles min^{-1}) and noted an increase in collagen and protein synthesis without a substantial increase in cell proliferation. The elastin membranes were prepared by slicing bovine aortic media and subjecting them to hot alkali to remove non-elastin components. This treatment yielded a sheet composed entirely of a network of elastin plates and interconnecting elastin fibres. Interestingly, although the study by Leung showed no increase in number of cells, the work by Sumpio & Banes (1988) showed inhibition of vsmc growth with their stretching regime. These discrepancies may be due to differences in the rate of stretching, the amount of strain generated, differences in the level of confluency of the cells, different substrate surfaces involved or a combination of these factors. Furthermore, endothelial cells subjected to a similar 3 cycles min^{-1} force regime as described above demonstrated an increase in DNA synthesis and cell number (Sumpio *et al.*, 1987) associated with striking changes in cytoskeletal orientation and protein electrophoretic patterns (Sumpio *et al.*, 1988b). Thus mechanical perturbation appears to be perceived differently by different cell types. Changes in collagen production by vsmc have been correlated with changes in cell structure (Sottiurai *et al.*, 1983). A four- to five-fold increase in rough endoplasmic reticulum and dilated cisternae was found depending on the duration of the stretching. The ability of an arterial wall to withstand the tensile forces associated with increases in blood pressure and to recover from mechanical injury requires cell proliferation as well as modifications in matrix composition and architecture of the media. There is evidence that the rate of connective tissue production

32   F. LYALL & M. R. DEEHAN

by smooth muscle cells is increased in hypertension and that this is closely related to the tensile strength imposed on the vessel wall (Wolinsky, 1970). The majority of smc of muscular and elastic arteries are orientated at an angle of 20–40° to the longitudinal axis of the artery. This corresponds to an angle of 50–70° between the longer axis of the smooth muscle cells and the resultant vector of the distending forces which lies at a right angle to the direction of blood flow. Betz, Roth & Schlote (1980) reported that repeated local transmural electrical stimulation of the carotid artery wall of rabbits causes a proliferation of smc into the subendothelial space. The smooth muscle cells of the proliferate lie with their longitudinal axis parallel to the direction of blood flow, i.e. at right angles to the resultant vector of the distending forces. Within 6 months after a 4-week stimulation period, the subendothelial cells of the proliferate change their orientation and are finally arranged as in the normal media, whereas the cells at the base of the proliferate remain in an orientation parallel to the longitudinal axis of the artery. This raised the question as to whether the change of orientation of the smc might be as a result of directional and rhythmic stretching of the artery wall by blood pulsation. Furthermore, perhaps mechanical forces were directing the arrangement of smc within the media. To address this question arterial smc from rabbit aortic media were grown on collagen-coated silicone membranes and then subjected to directional cyclic stretches and relaxations at a frequency of 50 times per minute (Dartsch, Hammerle & Betz, 1986). Cells which were stretched with an amplitude of 2% remained in random orientation after 14 days of continuously performed cyclic stretching. The cells which were stretched 5% for 12 days orientated at an angle of 61° to the direction of stretching while the cells which were stretched with an amplitude of 10% for 6 days orientated at an angle of 76°. The cells on the stationary and unstretched membranes remained in random orientation. These results suggested that stretching of the artery wall by blood pulsation may be a factor in determining the orientation of smc within the media of the artery wall and of those smc which proliferate into the subendothelial space after mechanical injury of the endothelium or electrical stimulation of the artery wall.

In summary, further studies are still required to understand fully the molecular and cellular mechanisms that control vascular changes in response to increased blood pressure and thus increased haemodynamic load. *In vitro* systems, which allow the direct effects of mechanical forces on cells to be investigated free from humoral influences, may help further our knowledge in this area.

References

Betz, E., Roth, J. & Schlote, W. (1980). Proliferation of smooth muscle cell in long-term local electrical stimulation of the carotid arteries. *Folia angiol* **28**, 27–31.

Bevan, R. D., Marthens, E. V. & Bevan, J. A. (1976). Hypertension of vascular smooth muscle in experimental hypertension in the rabbit. *Circulation Research* **38** (Suppl. II), 58–62.

Bevan, R. D. (1984). Trophic effects of peripheral adrenergic nerves on vascular structure. *Hypertension* **6** (Suppl. III); III-19–26.

Burke, J. M. & Ross, R. (1979). Synthesis of connective tissue molecules by smooth muscle. *International Reviews of Connective Tissue Research* **8**, 119–57.

Christensen, K. L. (1991). Reducing pulse pressure in hypertension may normalize small artery structure. *Hypertension* **18**, 722–7.

Dartsch, P. C., Hammerle, H. & Betz, E. (1986). Orientation of cultured arterial smooth muscle cells growing on cyclically stretched substrates. *Acta anat* **125**, 108–13.

Folkow, B. (1975). Vascular changes in hypertension – a review and recent animal studies. In *Pathophysiology and Management of Arterial Hypertension*, ed. G. Berglund, L. Hansson & L. Werkö, pp. 95–113. Mölndal, Sweden: L. Lidgren and Söner.

Folkow, B. (1978). The fourth Volhard Lecture. Cardiovascular structural adaptation: its role in initiation and maintenance of primary hypertension. *Clinical Science and Molecular Medicine* **55** (Suppl. 4), 3S–22S.

Freis, E. D. (1969). Hypertension and atherosclerotic disease. *American Journal of Medicine* **46**, 735–40.

Furuyama, M. (1962). Histometrical investigations of arteries in reference to arterial hypertension. *Tohoku Journal of Experimental Medicine* **76**, 388–414.

Izzard, A. S., MacIver, D. H., Cragoe, E. J. & Heagerty, A. M. (1991). Intracellular pH in rat resistance arteries during the development of experimental hypertension. *Clinical Science* **81**, 65–72.

Kanbe, T., Nara, Y., Tagami, M. & Yamori, Y. (1983). Studies of hypertension-induced vascular hypertrophy in cultured smooth muscle cells from spontaneously hypertensive rats. *Hypertension* **5**, 887–92.

Korsgaard, N. & Mulvany, M. J. (1988). Cellular hypertrophy in mesenteric resistance vessels from renal hypertensive rats. *Hypertension* **12**, 162–7.

Lee, R. M. K. W. (1985). Vascular changes at the prehypertensive phase in the mesenteric arteries from spontaneously hypertensive rats. *Blood Vessels* **22**, 105–26.

Lee, R. M. K. W., Garfield, R. E., Forrest, J. B. & Daniel, E. E. (1983). Morphometric study of structural changes in the mesenteric blood vessels of spontaneously hypertensive rats. *Blood Vessels* **20**, 57–71.

Lee, R. M. K. W. & Smeda, J. S. (1985). Primary versus secondary structural changes of the blood vessels in hypertension. *Canadian Journal of Physiology and Pharmacology* **63**, 392–401.

Lee, R. M. K. W. & Triggle, C. R. (1986). Morphometric study of mesenteric arteries from genetically hypertensive Dahl strain rats. *Blood Vessels* **23**, 199–224.

Leung, D. Y. M., Glagov, S. & Mathews, M. B. (1977). Elastin and collagen accumulation in rabbit ascending aorta and pulmonary trunk during post-natal growth: correlation of cellular synthetic response with medial tension. *Circulation Research* **41**, 316–324.

Lever, A. F. (1986). Slow pressor mechanisms in hypertension: a role for hypertrophy of resistance vessels. *Journal of Hypertension* **4**, 515–24.

Limas, C., Westrum, B. & Limas, C. J. (1980). The evolution of vascular changes in the spontaneously hypertensive rat. *American Journal of Pathology* **98**, 357–84.

Lyall, F., Dornan, E. S., McQueen, J., Boswell, F. & Kelly, M. R. (1992). Angiotensin II increases proto-oncogene expression and phosphoinositide turnover in vascular smooth muscle cells via the angiotensin II AT_1 receptor. *Journal of Hypertension* **10**, 1463–9.

MacIver, D. H., Green, N. K., Gammage, M. D., Durkin, H., Izzard, A. S., Franklyn, J. A. & Heagerty, A. M. (1993). Effect of experimental hypertension on phosphoinositide hydrolysis and proto-oncogene expression in cardiovascular tissues. *Journal of Vascular Research* **30**, 13–22.

Mayne, R. (1984). Vascular connective tissue: normal biology and derangement in human diseases. In *Diseases of Connective Tissue: The Molecular Pathology of the Extracellular Matrix*, ed. J. Uitto & Z. Perejda, pp. 271–308. New York: Marcel Dekker.

Mulvany, M. J. (1986). Role of vascular structure in blood pressure development of the spontaneously hypertensive rat. *Journal of Hypertension* **4** (Suppl. 3), S61–3.

Mulvany, M. J., Baandrup, U. & Gundersen, H. J. G. (1985). Evidence for hyperplasia in mesenteric resistance vessels of spontaneously hypertensive rats using a three-dimensional disector. *Circulation Research* **57**, 794–800.

Naftilan, A. J., Pratt, R. E., Eldridge, C. S., Lin, H. L. & Dzau, V. J. (1989). Angiotensin II induces c-*fos* expression in smooth muscle via transcriptional control. *Hypertension* **13**, 706–11.

Olivetti, G., Anversa, P., Melissari, M. & Loud, A. (1980). Morphometry of medial hypertrophy in the rat thoracic aorta. *Laboratory Investigation* **42**, 559–65.

Olivetti, G., Melissari, M., Marchetti, G. & Anversa, P. (1982). Quantitative structural changes of the rat thoracic aorta in early spontaneous hypertension: tissue composition and hypertrophy and hyperplasia of smooth muscle cells. *Circulation Research* **51**, 19–26.

Owens, G. K., Rabinovitch, P. S. & Schwartz, S. M. (1981). Smooth muscle cell hypertrophy versus hyperplasia in hypertension. *Proceedings of the National Academy of Sciences USA* **78**, 7759–63.

Owens, G. K. & Reidy, M. A. (1985). Hyperplastic growth response of vascular smooth muscle cells following induction of acute hypertension in rats by aortic coarctation. *Circulation Research* **57**, 695–705.

Owens, G. K. & Schwartz, S. M. (1982). Alterations in vascular smooth muscle mass in the spontaneously hypertensive rat: role of cellular hypertrophy, hyperploidy and hyperplasia. *Circulation Research* **51**, 280–9.

Owens, G. K. & Schwartz, S. M. (1983). Vascular smooth muscle cell hypertrophy and hyperploidy in Goldblatt hypertensive rat. *Circulation Research* **53**, 491–501.

Ross, R. (1981). Atherosclerosis: a problem of the biology of arterial wall cells and their interactions with blood components. *Arteriosclerosis* **1**, 293–311.

Ross, R. (1986). The pathogenesis of atherosclerosis – an update. *New England Journal of Medicine* **314**, 488.

Schwartz, S. M. & Ross, R. (1984). Cellular proliferation in atherosclerosis and hypertension. *Progress in Cardiovascular Diseases* **26**, 355–72.

Sottiurai, V. S., Kollros, P., Glagov, S., Zarins, C. K. & Mathews, M. B. (1983). *Journal of Surgical Research* **35**, 490–7.

Sumpio, B. E. & Banes, A. J. (1988). Response of porcine aortic smooth muscle cells to cyclic tensional deformation in culture. *Journal of Surgical Research* **44**, 696–701.

Sumpio, B. E., Banes, A. J., Buckley, M. & Johnson, G. (1988b). Alterations in aortic endothelial cell morphology and cytoskeletal protein synthesis during cyclic tensional deformation. *Journal of Vascular Surgery* **7**, 130–8.

Sumpio, B. E., Banes, A. J., Levin, L. G. & Johnson, G. (1987). Mechanical stress stimulates aortic endothelial cells to proliferate. *Journal of Vascular Surgery* **6**, 404–10.

Sumpio, B. E., Banes, A. J., Link, W. G. & Johnson, G. (1988a). Enhanced collagen production by smooth muscle cells during repetitive mechanical stretching. *Archives of Surgery* **123**, 1233–6.

Taubman, M. B., Bradford, C. B., Izumo, S., Tsuda, T., Alexander, R. W. & Nadal-Ginard, B. (1989). Ang II induces c-*fos* mRNA in

aortic smooth muscle. *Journal of Biological Chemistry* **264**, 526–30.
Wolinsky, H. (1970). Response of the rat aortic media to hypertension: morphological and chemical studies. *Circulation Research* **26**, 507–22.
Zak, R. (1974). Development and proliferative capacity of cardiac muscle cells. *Circulation Research* **35** (Suppl. II) II-17–26.

P. BODIN, A. LOESCH, P. MILNER
and G. BURNSTOCK

Effect of increased flow on release of vasoactive substances from vascular endothelial cells

Endothelium-derived vasoactive substances

In 1980, Furchgott and Zawadski established the role of vascular endo-thelium in mediating the relaxation of arterial smooth muscle to acetyl-choline (Furchgott & Zawadski, 1980). Many other substances, whose ability to initiate changes in vascular tone also depends on the integrity of the endothelium, were soon discovered. These substances include vasopressin (Katusic, Shepherd & Vanhoutte, 1984), bradykinin (Cherry et al., 1982), angiotensin II (Toda, 1984), ADP (De Mey & Vanhoutte, 1981), ATP (Burnstock & Kennedy, 1985), noradrenaline and serotonin (Cocks & Angus, 1983). In order to produce their effect, these substances bind to specific receptors located on the surface of endothelial cells and subsequently provoke the release of an endothelial factor which acts on the vascular smooth muscle cells. This diffusible factor, termed endo-thelium-derived relaxing factor (EDRF), with a very short biological half-life, was identified in 1987 as nitric oxide (Ignarro et al., 1987; Khan & Furchgott, 1987; Palmer, Ferrige & Moncada, 1987) or a related labile nitroso compound (Rubanyi et al., 1989). However, it has been recog-nised that not all endothelium-dependent vascular responses can be ex-plained by release of nitric oxide (Rubanyi & Vanhoutte, 1987). Prosta-cyclin is also released by endothelial cells and mediates vasodilation (Moncada et al., 1976). The existence of a substance mediating endothe-lium-dependent hyperpolarisation of smooth muscle has also been postu-lated (Komori & Suzuki, 1987). Soon after the discovery of EDRF, it was shown that endothelial cells could mediate vasoconstriction as well as vasodilatation (De Mey & Vanhoutte, 1982). These endothelium-dependent constrictions are mediated by diffusible substances whose exact nature has not been established. These substances are termed endo-thelium-derived contracting factors (EDCF). Isolation and purification

Society for Experimental Biology Seminar Series 54: *Biomechanics and Cells*, ed. F. Lyall & A. J. El Haj. © Cambridge University Press 1994, pp. 37–60.

of the 21-amino acid vasoconstrictor peptide, endothelin, from porcine aortic endothelial cells further emphasised the importance of the endothelium in modulating vascular tone (Yanagisawa *et al.*, 1988) even though endothelin is not likely to be the EDCF mediating rapid changes in vascular tone (Rubanyi, 1989).

As a result of these findings, a new concept for the control of vascular tone in mammalian blood vessels has emerged. Vascular smooth muscle is located between two regulators: the autonomic nervous system, which adjusts the vascular tone at a general level, and the endothelium, which adjusts the vascular tone at a local level (Lincoln & Burnstock, 1990; see Fig. 1).

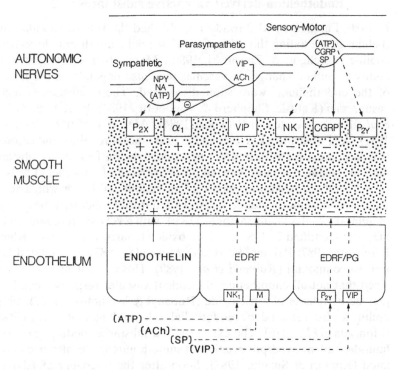

Fig. 1. A cartoon illustrating the regulation of the vascular tone by endothelial cells and neural mechanisms. Neuromediators are released by nerve endings to act directly on smooth muscle cells and cause either vasoconstriction or vasodilatation. Vasoactive substances act on endothelial cells and lead to the release of EDRF and/or prostacyclin, causing vasodilatation of the vascular wall. Endothelial cells also synthesise endothelin, able to induce contraction of the smooth muscle cells.

Local source of endothelium-dependent vasoactive substances

For a vasoactive substance to contribute to the regulation of vascular tone via an endothelium-dependent mechanism it must be present in the vicinity of the endothelial cells (Vanhoutte & Miller, 1985). Despite the fact that most of the substances initiating endothelium-dependent responses are also neurotransmitters, it is unlikely that perivascular nerves are their source since in most vessels (with the exception of microvessels), nerve terminations are exclusively located in the adventitia, posing several diffusion barriers between nerves and the luminal surface. This was demonstrated with acetylcholine which is unable to reach endothelial cells when released from perivascular nerves (Cohen, Shepherd & Vanhoutte, 1984). ATP and serotonin are released in the vicinity of endothelial cells by aggregating platelets. ATP is also released by red blood cells under hypoxic conditions (Bergfeld & Forester, 1992). However, blood structures are unlikely to be the source of acetylcholine, vasopressin or substance P for which the plasma concentration is low as they are rapidly degraded. It soon became evident that the source of these substances was the endothelial cells themselves.

Doubts were overcome when some enzymes responsible for the formation of vasoactive substances, such as angiotensin converting enzyme, the enzyme responsible for the cleavage of angiotensin I to angiotensin II, and choline acetyltransferase, the enzyme responsible for the synthesis of acetylcholine, were identified and localised in vascular endothelial cells (Ryan, 1984; Parnavelas, Kelly & Burnstock, 1985). Since then, many substances, such as vasopressin, serotonin, angiotensin II and substance P, have been localised in endothelial cells from different vascular beds (Burnstock *et al.*, 1988; Loesch & Burnstock, 1988; Lincoln, Loesch & Burnstock, 1990). ATP and endothelin-1 are synthesised by endothelial cells (Pearson *et al.*, 1978; Yanagisawa *et al.*, 1988). However, it is not yet established if all the vasoactive substances localised in endothelial cells are synthesised by these cells. It is well known that some of the substances circulating in the blood, such as serotonin, choline and adenosine, are taken up into endothelial cells by an active mechanism (Junod & Ody, 1977; Pearson *et al.*, 1978; Hamel *et al.*, 1987).

Studies both *in situ* and in cultures have shown that the population of endothelial cells is heterogeneous, with regard to their content of vasoactive substances. In this population, both immunopositive and immunonegative cells for one or another vasoactive substance can be seen (Loesch, Bodin & Burnstock, 1991; Cai *et al.*, 1993: Fig. 2). In the same

40   P. BODIN *et al.*

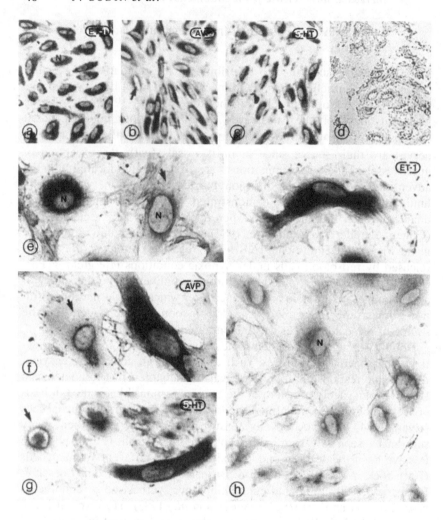

Fig. 2. Light microscopy of rabbit aortic endothelial cells in primary culture. Cells were labelled for endothelin-1 (*a,e*), vasopressin (*b,f*) and serotonin (*c,g*). Immunonegative cells can also be seen (arrows). (*e*) A montage of two immunopositive cells from the same preparation displaying a different pattern of staining. (*d*) Control by omission of the primary antibody. (*h*) Control by replacement of the antibody with normal rabbit serum. N, cell nucleus. (*a–d*) 230 ×, (*e–h*) 500 ×. (From Loesch *et al.*, 1991. Reproduced by permission of Pergamon Press.)

population, it is not uncommon to find several different substances co-localised inside the same endothelial cell (Loesch *et al.*, 1991).

In their location, at the luminal surface of blood vessels, in the intimal layer of the vascular wall, endothelial cells are ideally situated to sense changes in the circulating blood, such as variations of pressure, viscosity, flow rate, pH or chemical composition (Rubanyi, Romero & Vanhoutte, 1986; Harder, 1987). When mechanical forces such as shear stress, pressure or active stretch are applied to endothelial cells, the vessels undergo a large vasodilatation that is mediated by EDRF and abolished by removal of the endothelium (Kaiser, Hull & Sparks, 1986; Förstermann, Dudel & Frölich, 1987). One way of studying the role of vascular endothelial cells as a source of releasable vasoactive substances is thus to stimulate them mechanically. Increase of the perfusion flow rate is very easy to produce and does not necessitate a sophisticated set-up. This stimulus has therefore been used to investigate the release profile of endothelial cells from different vascular beds. The concentrations of vasoactive substances released in the perfusate overflowing endothelial cells have been determined using several techniques. Chemiluminescence has been used to measure the concentrations of ATP (Kirkpatrick & Burnstock, 1987) and acetylcholine (Israel & Lesbats, 1982). The concentrations of peptides have been measured using inhibition enzyme-linked immunosorbent assays (ELISA) (Milner *et al.*, 1989).

Release of vasoactive substances from vascular beds

In 1990, a study (Ralevic *et al.*, 1990) showed that the perfused rat hind-limb, *in vivo*, was able to release substance P when stimulated by increased flow. Introduction of air bubbles into the vasculature of the hind-limb, a procedure which removes endothelial cells from the arteries but not from arterioles and capillaries, suppressed the release of substance P. It was thus concluded that substance P released by the hind-limb during exposure to increased flow was unlikely to arise from perivascular nerves but from endothelial cells. This was confirmed by the fact that in animals treated with capsaicin, a neurotoxin depleting primary sensory neurones, the vessels released as much substance P as the control preparations.

This first series of experiments indicated that studying the release properties of vascular endothelial cells under conditions of increased flow would be beneficial in understanding the regulation of smooth muscle tone by the endothelium. However, this study also indicated that a complete characterisation of the release properties of endothelial cells would be quite difficult using whole animals or isolated organs as there are often changes in pressure and oedema formation, concomitant with increases in flow rate, that could influence the release of vasoactive substances. It transpired that the best technique to explore endothelial cell properties was to use isolated cells.

Release of vasoactive substances from cultured endothelial cells

Endothelial cells from human umbilical vein were cultured and passaged on microcarriers. Cells were perfused at 0.5 ml min^{-1} and stimulated twice for 3 min at 1.5 ml min^{-1}. Upon stimulation by increased flow, cells released ATP, substance P and acetylcholine. ATP was consistently rapidly released at the onset of increased flow rate (Fig. 3) whereas release of substance P and acetylcholine was slower and more varied. Substance P was released from only four out of 16 cultures of endothelial cells and acetylcholine from four out of 12 cultures (Milner *et al.*, 1990a).

Despite presenting certain advantages compared with the use of a whole vascular bed, the growth of cells on microcarrier beads necessitates a large initial number of endothelial cells, at least 2 weeks of culture and requires expensive equipment. Endothelial cells cultured on Petri dishes have often been used to study the release of endogenous substances into the culture medium in response to various stimuli (Castellot *et al.*, 1981; Emori *et al.*, 1989). Cells in each dish can be stimulated but comparison between dishes may be hazardous due to variations in the number of cells in each culture. In any case, the use of these techniques is limited as the cells cannot be perfused and freshly isolated cells cannot be used.

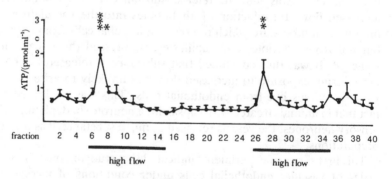

Fig. 3. Levels of ATP in fractions collected from endothelial cells isolated from human umbilical vein. Cells were perfused at 0.5 ml min^{-1} and stimulated at 1.5 ml min^{-1}.*** $p < 0.001$, ** $p < 0.002$, * $p < 0.01$. (From Milner *et al.*, 1990a. Reproduced by permission of the Royal Society, London.)

It is essential to use freshly isolated cells when comparing the characteristics of endothelial cells in pathological animal models as the cells, once cultured, could have lost some of the characteristics they developed in response to the pathology.

Release of vasoactive substances from freshly isolated cells

In order to study the release properties of freshly isolated endothelial cells, a novel experimental system has been developed. In this system, vascular endothelial cells are placed on untreated 3 μm pore size filter membranes made of mixed ester of cellulose (cellulose acetate and cellulose nitrate). Endothelial cells can be cultured on this substratum and grow as monolayers of closely apposed polygonal cells, show a cobblestone pattern at confluency and do not appear morphologically different from cells cultured on Petri dishes or coverslips coated with collagen or gelatin (Fig. 4). Electron microscopy of these cells shows numerous vesicular and membranous organelles indicating active synthetic and metabolic properties. Moreover, endothelial cells grown on a filter membrane are heterogeneous, as described in primary cultures of aortic endothelial cells of rabbit aorta and *in situ* (Loesch *et al.*, 1991). To study their release properties, freshly isolated or cultured endothelial cells on the cellulose membrane are placed in a filter holder. The filter is kept at 37 °C throughout the experiment. Using a peristaltic pump, endothelial cells are perfused with Krebs buffer at a low flow rate (0.5 ml min^{-1}) and stimulated twice for 3 min by an increased flow rate (3.0 ml min^{-1}) (see Fig. 5). The number of cells is estimated by determination of their protein content (Bradford, 1976).

Using this experimental system, it was shown that upon stimulation by increased flow, the increased release of ATP was specific to endothelial cells since vascular smooth muscle cells isolated from the same animals did not show an increase in release under the same experimental conditions (Bodin, Bailey & Burnstock, 1991; Fig. 6). The release of ATP by endothelial cells was not due to cell lysis since (i) the same cells could be stimulated several consecutive times by increased flow, (ii) there was no detectable lactate dehydrogenase activity in the perfusate, (iii) the cells did not appear morphologically different before and after perfusion.

The release of ATP, endothelin and arg-vasopressin from freshly isolated adult rabbit aortic endothelial cells subjected to short periods of increased perfusion flow rate was investigated (Milner *et al.*, 1990b: see Fig. 7). Endothelial cells released small amounts of these substances during perfusion at low flow rate. Stimulation by increased flow rate led to a rapid increase of the levels of endothelin and ATP. During the

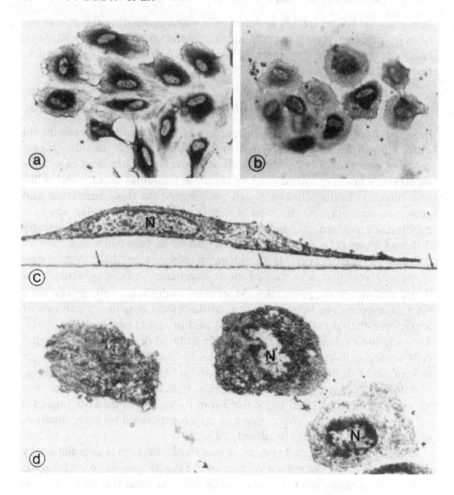

Fig. 4. (*a*) Group of endothelin-positive endothelial cells cultured on collagen-coated coverslips, 400 ×. (*b*) Endothelin-positive endothelial cells cultured on cellulose filter, 400 ×. No morphological difference is noticeable between the two preparations. (*c*) Electron microscopy of an endothelial cell attached to the filter. Note the surface of the filter membrane (arrows) 10500×. (*d*) Tangential section (*en face*) of endothelin-positive cells on filter, 8200×, N, cell nucleus.

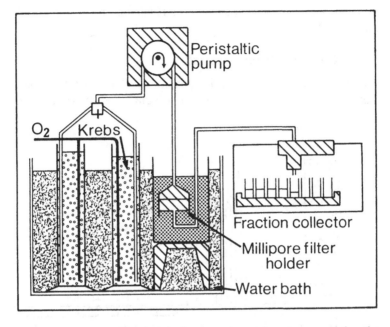

Fig. 5. A diagram illustrating the experimental system used for the study of the release properties of either cultured or freshly isolated endothelial cells. Cells are placed or grown on a 3 µm pore size cellulose membrane contained in a filter holder. Cells are perfused with oxygenated Krebs buffer at 37 °C.

second period of stimulation, the release of ATP was smaller while the release of endothelin was greater than during the first period. Arg-vasopressin remained at the basal level. The diminished release of ATP during the second period of stimulation may have been due to a limiting source of adenosine in the perfusion medium (Pearson & Gordon, 1979). It has been shown that shear stress leads to an increase in the expression of mRNA encoding endothelin (Yoshizumi *et al.*, 1989), which could explain the larger release of endothelin during the second stimulation. The fact that arg-vasopressin was not released upon stimulation re-inforces the idea of the specificity of the release of ATP and endothelin, especially since arg-vasopressin is often co-localised with endothelin in the same cells (Loesch *et al.*, 1991). At this stage, however, it was not clear whether there was a relationship between ATP release and endo-thelin release. The physiological implications of the simultaneous release of a vasodilator (ATP) with a potent vasoconstrictor (endothelin) was intriguing even though it is known that other stimuli, such as thrombin, cause release of these two substances (Pearson & Gordon, 1979; Schini *et al.*, 1989).

Effect of chronic hypoxia

In order to study the potential physiological role of the concomitant release of ATP and endothelin, experiments were made comparing the

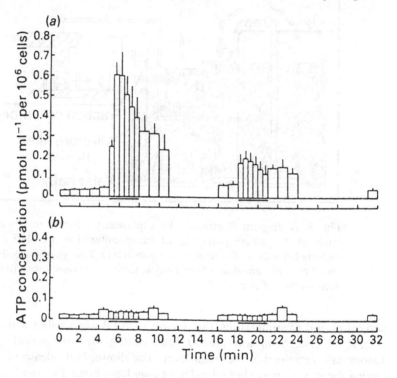

Fig. 6. ATP release from (*a*) endothelial cells and (*b*) smooth muscle cells freshly isolated from rabbit thoracic aorta. Heavy lines (——) represent periods of increased flow from 0.5 to 3.0 ml min⁻¹. Results are expressed as picomoles of ATP released per ml and per million cells. (From Bodin *et al.*, 1991. Reproduced by permission of Macmillan Press Ltd.)

Fig. 7. Release of (*a*) endothelin, (*b*) ATP, (*c*) vasopressin from freshly isolated rabbit aortic endothelial cells. Cells were perfused at a flow rate of 0.5 ml min⁻¹ and stimulated twice for 3 min at a flow rate of 3.0 ml min⁻¹. Heavy lines (——) represent periods of increased flow. (From Milner *et al.*, 1990b. Reproduced by permission of Academic Press, Inc.)

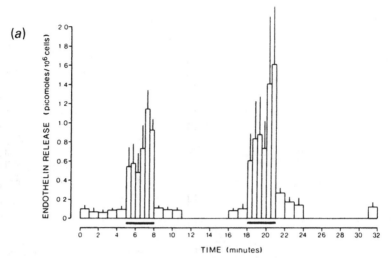

(a)

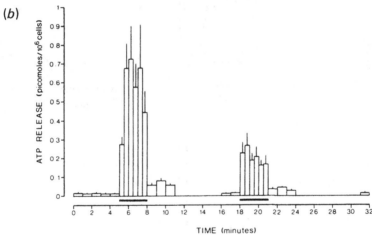

(b)

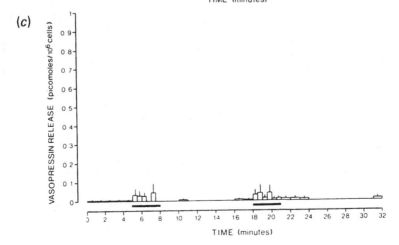

(c)

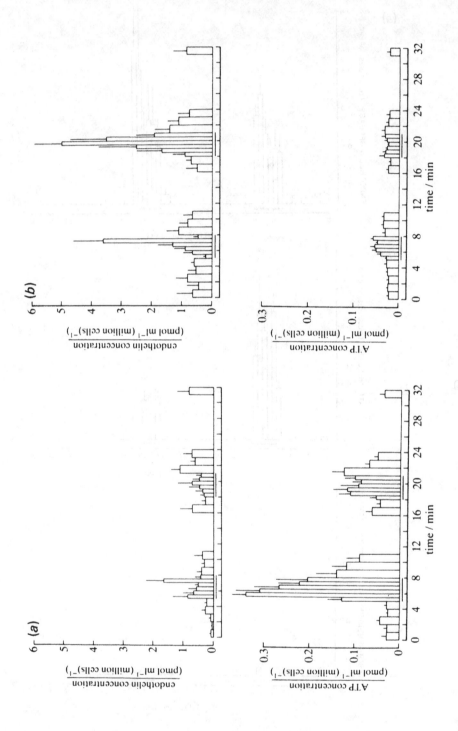

release profile from aortic endothelial cells isolated from normoxic rats and chronically hypoxic rats maintained in a hypoxic chamber for 10 days (Bodin *et al.*, 1992; Fig. 8). It was then shown that during periods of increased flow rate, endothelial cells isolated from normoxic rats increased their release of ATP and endothelin, as previously described in endothelial cells from the rabbit aorta (Milner *et al.*, 1990b). In comparison, in hypoxic rats, ATP release during stimulation was less, whereas endothelin release was greater than in normoxic controls. In both groups of animals, the release of vasopressin was not increased during periods of increased flow rate. It was concluded that the vessels exposed to reduced arterial oxygen tension might not respond to high flow stimulation by increasing their relaxation. There may be a dynamic balance between the release of ATP and endothelin so that in chronic hypoxia changes in endothelial cells occur which oppose the normal effect of increased flow rate.

Influence of age

A recent study has demonstrated that the rapid release of ATP and endothelin as seen in endothelial cells isolated from the aorta of 12-month-old rabbits does not occur in endothelial cells from 4-month-old rabbits (Fig. 9). Immunochemical studies have shown that the number of endothelial cells immunopositive for endothelin and arg-vasopressin in the rabbit aorta almost doubles between 4 and 12 months of age. Reduced endothelin availability may therefore account for the difference in quantities of endothelin released with an increased flow rate between the two age groups (Milner *et al.*, 1992).

The simultaneous release of a vasoconstrictor and a vasodilator agent by endothelial cells could reflect their ability to adjust the response of the vessel, achieving a fine control of the vascular tone. It is thus believed that under physiological conditions, a balance in the release of EDRF and EDCF contributes to the maintenance of vascular tone. In this way, in normal conditions, endothelin and ATP released during periods of increased flow would be a normal part of the response to the stimulus.

Moreover, ATP is not the only vasodilator agent released by endo-

---

Fig. 8. Release of (*a*) endothelin, (*b*) ATP by freshly isolated endothelial cells from the thoracic aorta of normoxic and chronically hypoxic rats. Cells were perfused at 0.5 ml min^{-1} and stimulated twice for 3 min at 3.0 ml min^{-1}. Heavy lines (———) represent periods of increased flow. (From Bodin *et al.*, 1992. Reproduced by permission of the Royal Society.)

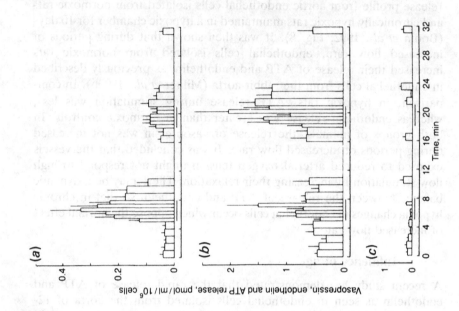

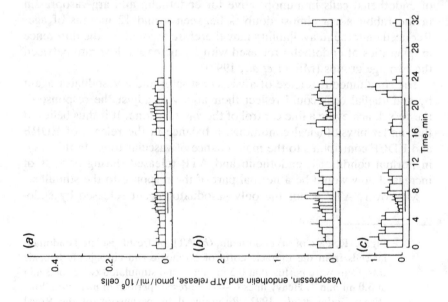

thelial cells upon stimulation. Substance P, vasopressin and ATP have already been shown to be simultaneously released upon stimulation of the cerebral vascular bed by increased flow (Domer *et al.*, 1992). Similarly, endothelin is probably not the only vasoconstrictor agent released by endothelial cells during stimulation. Thus, it is likely that endothelial cells release simultaneously several vasoconstrictor and vasodilator agents in response to stimulation. This will need to be incorporated into the overall picture of local control mechanisms when more information is available.

When vascular endothelial cells are stimulated by increases in flow rate, they rapidly release specific vasoactive substances. Electron microscopic observations of vascular endothelial cells have shown that immunoreactivity to vasoactive substances such as vasopressin, serotonin and endothelin is not confined to any particular subcellular structures. Instead, this immunoreactivity is widespread in the cytoplasmic matrix, perhaps associated with free ribosomes (Loesch *et al.*, 1991: see Fig. 10). It has already been shown that the release of endothelin is constitutive (i.e. from the cytoplasmic phase) as opposed to regulated (i.e. from the vesicular phase) (Yanagisawa *et al.*, 1988). Considering their location in the cell cytoplasm, it is therefore likely that the rapid release of other vasoactive substances from endothelial cells is also constitutive.

Mechanisms of release of vasoactive substances

Until now, the mechanisms of release of vasoactive substances by vascular endothelial cells have not been determined. However, stretch or shear stress activated K^+ ionic currents might be involved (Olesen, Clapham & Davies, 1988). There is also clear evidence that elevation of the concentration of intracellular Ca^{2+} is required for the release of EDRF (Long & Stone, 1985; Lewis & Smith, 1991). An increase of the intracellular concentration of Ca^{2+} has also been reported in response to shear stress (Rubanyi *et al.*, 1986). Although the nature and the role of the K^+ conductance has not been elucidated, it has been proposed that the efflux of K^+ and hyperpolarisation of the endothelial cells would produce a rise in the concentration of intracellular Ca^{2+}, necessary for the release of EDRF (Busse, Luckhoff & Pohl, 1992). Using the pulmonary vascular

---

Fig. 9. Release of (*a*) ATP, (*b*) endothelin, (*c*) vasopressin from freshly isolated endothelial cells isolated from 4- and 12-month-old rabbits. Cells were perfused at 0.5 ml min^{-1} and stimulated twice for 3 min at 3.0 ml min^{-1}. Heavy lines (——) represent periods of increased flow. (From Milner *et al.*, 1992. Reproduced by permission of S. Karger.)

52   P. BODIN *et al.*

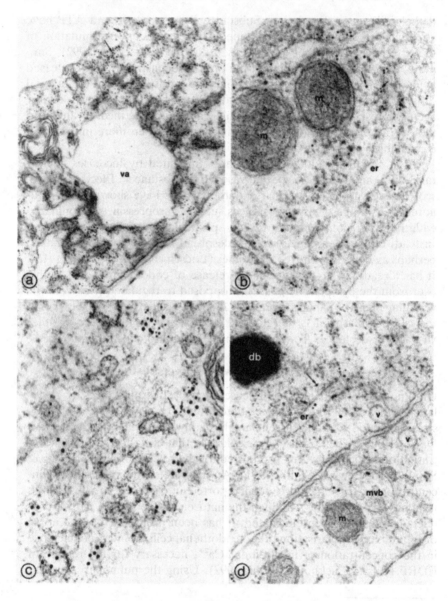

Fig. 10. Electron micrographs of endothelial cell profiles labelled for
two different antigenic sites by gold particles of 10 nm (long arrows) and
5 nm (short arrows). (*a*) A cell profile showing abundant cytoplasmic
colocalisation of serotonin (long arrow) and endothelin (short arrow);

bed of the rat, a recent study looked at some biochemical events involved in the release of ATP by increased flow rate (Hassessian, Bodin & Burnstock, 1993). This work showed that increases in flow produced proportional increases in vascular perfusion pressure. In the presence of suramin, used as a P_2 purinoceptor antagonist, the perfusion pressure was significantly higher than in the control group, indicating that ATP released into the lumen functions to dilate the pulmonary bed. In addition, the effect of suramin was greater, the higher the flow rate suggesting a vasodilator role of ATP increasing with shear stress. ATP was also released by the perfused lung preparation stimulated by increased flow. Glybenclamide, a blocker of intracellular ATP-sensitive K^+ channels (Cook, 1988) increased the vascular resistance under conditions of increased flow rate. At the same time, glybenclamide was found to block ATP release in response to increased flow rate (Fig. 11).

Conclusion

It is likely that stimulation of a vascular bed by mechanical forces such as increased flow produces a succession of changes in endothelial cells such as increase in mRNA production (Yoshizumi *et al.*, 1989), opening of stretch-activated and/or intracellular ATP-sensitive K^+ ionic channels (Olesen *et al.*, 1988; Hassessian *et al.*, 1993) and increase in the intracellular concentration of Ca^{2+}, leading to the rapid release of vasoactive substances into the blood (Fig. 12). These substances can then act on specific receptors located on endothelial cells which leads to the release of EDRF, EDCF and/or prostacyclin, acting in turn on smooth muscle cells.

Assessment of the release profiles of endothelial cells from different vascular beds subjected to increased flow reveals a heterogeneity. For instance, vasopressin is released by the perfused rabbit cerebral vasculature but not by the other vascular beds studied (Domer *et al.*, 1992).

---

va, vacuole. (*b*) A cell profile labelled for vasopressin (long arrow) and serotonin (short arrow) displaying gold particles near the cisternae of granular endoplasmic reticulum (er); m, mitochondria. (*c*) A cell profile displaying cytoplasmic labelling for endothelin (long arrow) and vasopressin (short arrow). (*d*) Two cell profiles labelled for endothelin (long arrow) and vasopressin (short arrow). Gold particles are present in the cytoplasm and in association with granular endoplasmic reticulum. Subplasmalemmal vesicles (v) are free of gold particles; db, dense body; mvb, multivesicular body. (*a*) 50 000×; (*b*) 120 000×; (*c*) 110 000×; (*d*) 100 000×. (From Loesch *et al.*, 1991. Reproduced by permission of Pergamon Press.)

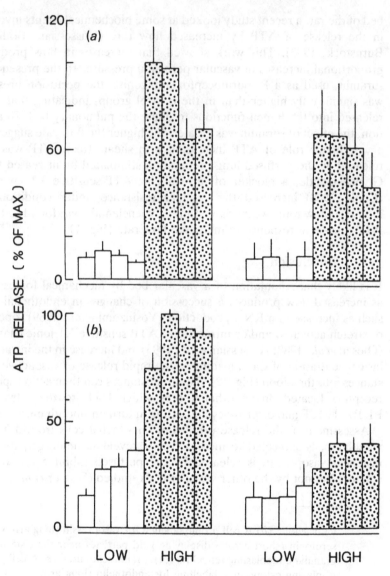

Fig. 11. Release of ATP by perfused rat lung preparation. The flow rate was increased twice from 8.4 to 27.3 ml min^{-1} (stippled bars), 30 min apart. (*a*) The release of ATP was significantly increased during both periods of high flow rate. In the presence of glybenclamide (*b*), the release of ATP was significantly reduced during the second step of augmentation of flow rate. (From Hassessian *et al.*, 1993. Reproduced by permission of Macmillan Press Ltd.)

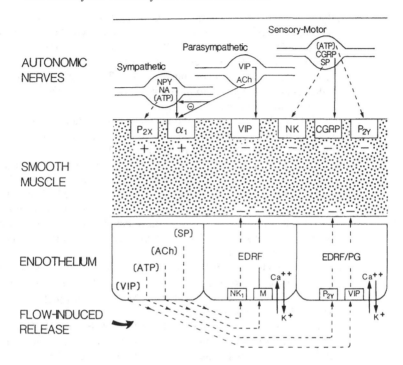

Fig. 12. A cartoon illustrating the regulation of the vascular tone by endothelial cells and neural mechanisms. Upon stimulation by increased flow, K^+ and Ca^{2+} ionic currents are activated. Vasoactive substances are selectively released by endothelial cells. These vasoactive substances act on endothelial cells and lead to the release of EDRF and/or prostacyclin, causing vasodilatation of the vascular wall.

Results presented in this chapter show that endothelial cells are the source, as well as the target, of vasoactive substances. The release of these vasoactive substances could play a physiological role in the overall control of vascular tone by endothelial cells. Furthermore, changes in the release properties of the cells could be implied in pathological processes of the cardiovascular system.

References

Bergfeld, G. & Forester, T. (1992). Release of ATP from human erythrocytes in response to a brief pulse of hypoxia and hypercapnia. *Cardiovascular Research* **26**, 40–7.

Bodin, P., Bailey, D. & Burnstock, G. (1991). Increased flow-induced ATP release from isolated vascular endothelial cells but not smooth muscle cells. *British Journal of Pharmacology* **103**, 1203–5.

Bodin, P., Milner, P., Winter, R. & Burnstock, G. (1992). Chronic hypoxia changes the ratio of endothelin to ATP release from rat aortic endothelial cells exposed to high flow. *Proceedings of the Royal Society of London (Biol)* **247**, 131–5.

Bradford, M. M. (1976). A rapid and sensitive method for the quantification of microgram quantities of protein using the principle of protein dye binding. *Analytical Biochemistry* **72**, 248–54.

Burnstock, G. & Kennedy, C. (1985). Is there a basis for distinguishing two types of P_2-purinoceptor? *General Pharmacology* **16**, 433–40.

Burnstock, G., Lincoln, J., Fehér, E., Hopwood, A. M., Kirkpatrick, K., Milner, P. & Ralevic, V. (1988). Serotonin is localised in endothelial cells of coronary arteries and released during hypoxia: a possible new mechanism for hypoxia-induced vasodilation of the rat heart. *Experientia* **44**, 705–7.

Busse, R., Luckhoff, A. & Pohl, U. (1992). Role of membrane potential in the synthesis and release of EDRF. In *Endothelial Regulation of Vascular Tone*, ed. U. S. Ryan & G. M. Rubanyi, pp. 105–20. New York: Marcel Dekker.

Cai, W. Q., Bodin, P., Sexton, A., Loesch, A. & Burnstock, G. (1993). Localization of Neuropeptide Y and atrial natriuretic peptide in the endothelial cells of human umbilical blood vessels. *Cell and Tissue Research* **272**, 175–81.

Castellot, J. J., Jr, Addonizio, M. L., Rosenberg, R. & Karnovsky, M. J. (1981). Cultured endothelial cells produce a heparin-like inhibitor of smooth muscle cell growth. *Journal of Cell Biology* **90**, 372–9.

Cherry, P. D., Furchgott, R. F., Zawadski, J. V. & Jothianandan, D. (1982). Role of endothelial cells in the relaxation of isolated arteries by bradykinin. *Proceedings of the National Academy of Sciences USA* **79**, 2106–10.

Cocks, T. M. & Angus, J. A. (1983). Endothelium-dependent relaxation of coronary artery by noradrenaline and serotonin. *Nature* **305**, 627–30.

Cohen, R. A., Shepherd, J. T. & Vanhoutte, P. M. (1984). Neurogenic cholinergic prejunctional inhibition of sympathetic β-adrenergic relaxation in the canine coronary artery. *Journal of Pharmacology and Experimental Therapeutics* **229**, 417–21.

Cook, N. S. (1988). The pharmacology of potassium channels and their therapeutic potential. *Trends in Pharmacological Sciences* **8**, 209–11.

De Mey, J. G. & Vanhoutte, P. M. (1981). Role of the intima in cholinergic and purinergic relaxation of isolated canine femoral arteries. *Journal of Physiology (London)* **316**, 347–55.

De Mey, J. G. & Vanhoutte, P. M. (1982). Heterogeneous behaviour of the canine arterial and venous wall. *Circulation Research* 51, 439–47.

Domer, F., Alexander, B., Milner, P., Bodin, P. & Burnstock, G. (1992). Cerebrovascular perfusion of the rabbit brain: a method that permits evaluation of compounds in effluent samples. *Journal of Physiology (London)* 452, 323p.

Emori, T., Hirata, Y., Ohta, K., Shichiri, M. & Marumo, F. (1989). Secretory mechanism of immunoreactive endothelin in cultured bovine endothelial cells. *Biochemical and Biophysical Research Communications* 160, 93–100.

Förstermann, U., Dudel, C. & Frölich, J. C. (1987). EDRF is likely to modulate the tone of resistance arteries in rabbit hind-limb *in vivo*. *Journal of Pharmacology and Experimental Therapeutics* 243, 1055–61.

Furchgott, R. F. & Zawadski, J. V. (1980). The obligatory role of endothelial cells in the relaxation of arterial smooth muscle by acetylcholine. *Nature* 288, 373–6.

Hamel, E., Assumel-Lurdin, C., Edvinsson, L., Fage, D. & MacKenzie, E. T. (1987). Neuronal versus endothelial origin of vasoactive acetylcholine in pial vessels. *Brain Research* 420, 391–6.

Harder, D. R. (1987). Pressure-induced myogenic activation of cat cerebral arteries is dependent on intact endothelium. *Circulation Research* 60, 102–7.

Hassessian, H., Bodin, P. & Burnstock, G. (1993). Blockade by glibenclamide of the flow-induced endothelial release of ATP that contributes to the vasodilatation in the pulmonary vascular bed of the rat. *British Journal of Pharmacology* 109, 466–72.

Ignarro, L. J., Byrns, R. E., Bugas, G. M. & Woods, K. S. (1987). Endothelium-derived relaxing factor from pulmonary artery and vein possesses pharmacologic and chemical properties identical to those of nitric oxide radical. *Circulation Research* 61, 866–79.

Israel, M. & Lesbats, B. (1982). Application to mammalian tissues of the chemiluminescent method for detecting acetylcholine. *Journal of Neurochemistry* 39, 248–50.

Junod, A. F. & Ody, C. (1977). Amine uptake and metabolism by endothelium of pig pulmonary artery and aorta. *American Journal of Physiology* 232, C88–94.

Kaiser, L., Hull, S. S., Jr & Sparks, H. V., Jr (1986). Methylene blue and ETYA block flow-dependent dilation in canine femoral artery. *American Journal of Physiology* 250, H974–81.

Katusic, Z. S., Shepherd, J. T. & Vanhoutte, P. M. (1984). Vasopressin causes vascular dependent relaxations of the canine basilar artery. *Circulation Research* 55, 575–9.

Khan, M. T. & Furchgott, R. F. (1987). Similarities of behavior of nitric oxide and endothelium derived relaxing factor in a perfusion cascade bioassay system. *Federation Proceedings* 46, 385.

Kirkpatrick, K. & Burnstock, G. (1987). Sympathetic nerve-mediated release of ATP from the guinea-pig vas deferens is unaffected by reserpine. *European Journal of Pharmacology* **138**, 207–14.

Komori, K. & Suzuki, H. (1987). Heterogeneous distribution of muscarinic receptors in the rabbit saphenous artery. *British Journal of Pharmacology* **92**, 657–64.

Lewis, M. J. & Smith, J. A. (1991). Factors regulating the release of endothelium-derived relaxing factor. In *Endothelial Regulation of Vascular Tone*, ed. U. S. Ryan & G. M. Rubanyi, pp. 139–54. New York: Marcel Dekker.

Lincoln, J. & Burnstock, G. (1990). Neural–endothelial interactions in control of local blood flow. In *The Endothelium: An Introduction to Current Research*, ed. J. Warren, pp. 21–32. New York: Wiley–Liss.

Lincoln, J., Loesch, A. & Burnstock, G. (1990). Localization of vasopressin, serotonin and angiotensin II in endothelial cells of the renal and mesenteric arteries of the rat. *Cell and Tissue Research* **259**, 341–4.

Loesch, A., Bodin, P. & Burnstock, G. (1991). Colocalisation of endothelin, vasopressin and serotonin in cultured endothelial cells of rabbit aorta. *Peptides* **12**, 1095–103.

Loesch, A. & Burnstock, G. (1988). Ultrastructural localization of serotonin and substance P in vascular endothelial cells of rat femoral and mesenteric arteries. *Anatomy and Embryology* **178**, 137–42.

Long, C. J. & Stone, T. W. (1985). The release of endothelium-derived relaxant factor is calcium dependent. *Blood Vessels* **22**, 205–8.

Milner, P., Bodin, P., Loesch, A. & Burnstock, G. (1990b). Rapid release of endothelin and ATP from isolated aortic endothelial cells exposed to increased flow. *Biochemical and Biophysical Research Communications* **170**, 649–56.

Milner, P., Bodin, P., Loesch, A. & Burnstock, G. (1992). Increased shear stress leads to differential release of endothelin and ATP from isolated endothelial cells from 4- and 12-month-old male rabbit aorta. *Journal of Vascular Research* **29**, 420–5.

Milner, P., Kirkpatrick, K. A., Ralevic, V., Toothill, V., Pearson, J. & Burnstock, G. (1990a). Endothelial cells cultured from human umbilical vein release ATP, substance P and acetylcholine in response to increased flow. *Proceedings of the Royal Society of London (Biol)* **241**, 245–8.

Milner, P., Ralevic, V., Hopwood, A. M., Fehér, E., Lincoln, J., Kirkpatrick, K. & Burnstock, G. (1989). Ultrastructural localization of substance P and choline acetyltransferase. *Experientia* **45**, 121–5.

Moncada, S., Gryglewski, R., Bunting, S. & Vane, J. R. (1976). An enzyme isolated from arteries transforms prostaglandin endoperoxidase to an unstable substance that inhibits platelet aggregation. *Nature* **263**, 663–5.

Olesen, S. P., Clapham, D. E. & Davies, P. F. (1988). Haemodynamic shear stress activates a K^+ current in vascular endothelial cells. *Nature* **331**, 168–70.

Palmer, R. M. J., Ferrige, A. G. & Moncada, S. (1987). Nitric oxide release accounts for the biological activity of endothelium-derived relaxing factor. *Nature* **327**, 524–6.

Parnavelas, J. G., Kelly, W. & Burnstock, G. (1985). Ultrastructural localization of acetyltransferase in vascular endothelial cells in rat brain. *Nature* **316**, 724–5.

Pearson, J. D., Carleton, J. S., Hutchings, A. & Gordon, J. L. (1978). Uptake and metabolism of adenosine by pig aortic endothelial and smooth muscle cells in culture. *Biochemical Journal* **170**, 265–71.

Pearson, J. D. & Gordon, J. L. (1979). Vascular endothelial and smooth muscle cells in culture selectively release adenine nucleotide. *Nature* **281**, 384–6.

Ralevic, V., Milner, P., Hudlická, O., Kristek, F. & Burnstock, G. (1990). Substance P is released from the endothelium of normal and capsaicin-treated rat hind-limb vasculature, *in vivo*, by increased flow. *Circulation Research* **66**, 1178–83.

Rubanyi, G. M. (1989). Maintenance of 'basal' vascular tone may represent a physiologic role for endothelin. *Journal of Vascular Medicine and Biology* **1**, 315–16.

Rubanyi, G. M., Johns, A., Harrison, D. G. & Wilcox, D. (1989). Evidence that EDRF may be identical with an S-nitrosothiol and not with free nitric oxide. *Circulation* **80**, (Suppl. II), II-281.

Rubanyi, G. M., Romero, J. C. & Vanhoutte, P. M. (1986). Flow-induced release of endothelium-derived relaxing factor. *American Journal of Physiology* **250**, H1145–9.

Rubanyi, G. M. & Vanhoutte, P. M. (1987). Nature of endothelium-derived relaxing factor: are there two relaxing mediators. *Circulation Research* **61** (Suppl. II), 1161–7.

Ryan, J. W. (1984). The metabolism of angiotensin and bradykinin by endothelial cells. In *Biology of Endothelial Cells*, ed. E. A. Jaffe, pp. 317–29. Boston: Martinus Nijhoff.

Schini, V. B., Hendrickson, H., Heublein, D. M., Burnett, J. C., Jr & Vanhoutte, P. M. (1989). Thrombin enhances the release of endothelin from cultured porcine aortic endothelial cells. *European Journal of Pharmacology* **165**, 333–4.

Toda, N. (1984). Endothelium-dependent relaxation induced by angiotensin II and histamine in isolated arteries of the dog. *British Journal of Pharmacology* **80**, 301–7.

Vanhoutte, P. M. & Miller, V. M. (1985). Heterogeneity of endothelium-dependent responses in mammalian blood vessels. *Journal of Cardiovascular Pharmacology* **7** (Suppl. 3), S12–23.

Yanagisawa, M., Kurihara, H., Kimura, A., Tomobe, Y., Kabayashi,

M., Mitsui, Y., Yasaki, Y., Goto, K. & Masaki, T. (1988). A novel potent vasoconstrictor peptide produced by vascular endothelial cells. *Nature* 332, 411–15.

Yoshizumi, M., Kurihara, H., Sugiyama, T., Takaku, F., Yanagisawa, M., Masaki, T. & Yasaki, Y. (1989). Hemodynamic shear stress stimulated endothelin production by cultured endothelial cells. *Biochemical and Biophysical Research Communications* 161, 859–64.

T. M. GRIFFITH and I. R. HUTCHESON

Modulation of endothelium-derived relaxing factor activity by flow

Introduction

The discovery of endothelium-dependent relaxation by Furchgott and Zawadski in 1980 has provided major insights into the mechanisms through which circulating humoral agents, adventitial nerves, and the mechanical forces which result from blood flow influence vasomotor tone. The phenomenon is mediated by a labile endogeneous nitrovasodilator, endothelium-derived relaxing factor (EDRF), which can be detected electrochemically near the endothelial membrane as the nitric oxide radical (NO) (Malinski & Taha, 1992). A constitutive Ca^{2+}/calmodulin-dependent NO synthase (cNOS), which synthesises NO from a terminal guanidino nitrogen atom of L-arginine by incorporating molecular oxygen into both NO and L-citrulline, is in fact present under normal physiological circumstances in a variety of cell types (Moncada, Palmer & Higgs, 1991). Endothelial cNOS is only $c.$ 50% homologous to rat cerebellar cNOS, its amino terminus possessing structural features which probably confer the ability to respond to mechanical signals resulting from flow (Nishida et al., 1992). These include sites for proline-directed phosphorylation (e.g. by protein kinase A) and a potential substrate site for acyl transferase which may permit attachment to fatty acids (particularly myristic), and thus explain the essentially particulate nature of the enzyme.

An inducible NO synthase (iNOS), whose formation is suppressed by glucocorticoids and other protein synthesis inhibitors, is also expressed in endothelial, vascular smooth muscle and other cell types such as macrophages following immunological stimulation by endotoxin and certain cytokines (Joulou-Schaeffer et al., 1990; Moncada et al., 1991). This contrasts with endothelial cNOS which is down-regulated by exposure to cytokines (Nishida et al., 1992). iNOS is cytosolic and $c.$ 50% homo-

Society for Experimental Biology Seminar Series 54: *Biomechanics and Cells*, ed. F. Lyall & A. J. El Haj. © Cambridge University Press 1994, pp. 61–80.

logous to cNOS but does not require the Ca^{2+}/calmodulin complex for activation (Xie et al., 1992). It produces much larger amounts of NO than cNOS (nanomoles vs picomoles), thus contributing to the anergic responses of blood vessels to constrictor agonists in septic shock.

The principal mode of action of nitrovasodilators such as EDRF is to activate the cytosolic (soluble) guanylyl cyclase enzyme, thereby elevating cyclic guanosine monophosphate (cGMP) levels and facilitating phosphorylation of a wide range of proteins by cyclic GMP-dependent protein kinase (Fiscus, Rapoport & Murad, 1984). Participating mechanisms include: (i) phosphorylation and inhibition of myosin light chain kinase and down-regulation of myofilament Ca^{2+} sensitivity (Nishikawa et al., 1984; Nishimura & Van Breemen, 1989), (ii) stimulation of a plasmalemmal Ca^{2+}-extrusion ATPase (Popescu et al., 1985), (iii) depression of agonist-stimulated phosphoinositide turnover through reduced G protein activation, uncoupling of activated G protein to phospholipase C and smooth muscle hyperpolarisation (Rapoport, 1986; Hirata et al., 1990; Tare et al., 1990; Itoh et al., 1992), (iv) increased sequestration of cytosolic Ca^{2+} in sarcoplasmic reticulum by stimulation of a Ca^{2+}-ATPase (Twort & Van Breemen, 1988), (v) depression of Ca^{2+} influx through receptor-operated channels (Collins et al., 1986) which are inhibited by cGMP-dependent protein kinase (Blayney et al., 1991), and membrane hyperpolarisation which closes voltage-operated Ca^{2+} channels (Tare et al., 1990).

EDRF and flow-dependent dilatation

It has been recognised for many years that an increase in blood flow increases arterial calibre through a purely local mechanism (Schretzenmayr, 1933; Hilton, 1959). In conduit arteries this flow-dependent dilatation is endothelium-dependent, insensitive to cyclo-oxygenase inhibition (thus excluding a significant role for prostacyclin), activated within seconds, and maximal by c. 2 min in both animal models and human subjects (Holtz et al., 1984; Sinoway et al., 1989; Melkumyants & Balashov, 1990). There is now persuasive evidence that it is mediated by EDRF release in response to the mechanical stimulus of shear stress and is therefore sensitive to both the velocity of flow and perfusate viscosity (Pohl et al., 1986; Rubanyi, Romero & Vanhoutte, 1986; Tesfamarium & Cohen, 1988).

Cellular mechanisms mediating the EDRF response to flow

Endothelial cNOS is highly sensitive to changes in $[Ca^{2+}]_i$, and both prolonged agonist-induced and flow-related EDRF synthesis have an abso-

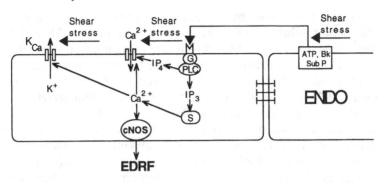

Fig. 1. The constitutive endothelial NO synthase is Ca^{2+}/calmodulin-dependent so that endothelium-dependent dilators and flow must both elevate $[Ca^{2+}]_i$ to increase EDRF synthesis. Both stimuli initiate an agonist/G protein/phospholipase C interaction which stimulates IP_3-evoked Ca^{2+} release from internal stores, and IP_4 may enhance influx of extracellular Ca^{2+} through synergism with $[Ca^{2+}]_i$. Shear stress activates a Ca^{2+}-dependent hyperpolarising K^+ current which increases the driving force for Ca^{2+} entry as endothelial cells do not possess voltage-operated Ca^{2+} channels, and also opens Ca^{2+} channels directly. Flow also stimulates release of endothelium-dependent agonists (e.g. ATP, bradykinin and substance P) from endothelial stores which may then act on neighbouring cells.

lute requirement for the presence of extracellular Ca^{2+}, and thus by implication Ca^{2+} influx (Fig. 1) (Griffith *et al.*, 1986; MacArthur *et al.*, 1993). Flow induces an initial transient rise in endothelial $[Ca^{2+}]_i$ followed by a plateau which is dependent on Ca^{2+} influx, these responses involving distinct pools of Ca^{2+} near the cell nucleus and near the cell membrane (Geiger *et al.*, 1992). The transient may involve mobilisation of Ca^{2+} from intracellular stores by inositol 1,4,5-trisphosphate (IP_3), whose formation is enhanced by steady shear stress and cyclical stretch (Nollert, Eskin & McIntire, 1990; Rosales & Sumpio, 1992). Guanine nucleotide regulatory proteins (G proteins), which are able to undergo a conformational change to modulate subcellular effector systems such as enzymes and ion channels, participate in this pathway (Miller & Burnett, 1992; Ohno *et al.*, 1992a). Synergism between IP_3 and IP_4 (inositol 1,3,4,5-tetrakisphosphate) may also increase the open state probability of $[Ca^{2+}]_i$-sensitive, Ca^{2+}-permeable membrane channels (Lückhoff & Clapham, 1992).

Additional mechanisms contribute to sustained Ca^{2+} influx, which cannot be mediated via voltage-operated channels as endothelial cells are electrically non-excitable (Lückhoff & Busse, 1986). Thus, shear stress stimulates Ca^{2+} influx by opening lanthanum-sensitive channels which also permit influx of other divalent cations (Lansman, Hallam & Rink, 1987; Schwartz et al., 1992). It also activates an inwardly rectifying outward K^+ current which causes membrane hyperpolarisation, thereby presumably increasing the driving potential for Ca^{2+} entry (Nakache & Gaub, 1988; Olesen, Clapham & Davies, 1988). This effect desensitises slowly, recovers on cessation of flow, and is fully activated at levels of shear stress within the in vivo range. Shear-induced release of EDRF-dependent vasodilators such as ATP, bradykinin, substance P and angiotensin II from endothelial stores also potentially contributes to flow-dependent dilatation (Figs. 1 and 2), there being experimental evidence for such a mechanism in respect of substance P and ATP (Ralevic et al., 1990, 1992; Bodin, Bailey & Burnstock, 1991). Furthermore, angiotensin converting enzyme inhibitors amplify basal EDRF activity by unmasking an autocrine effect of endogeneous bradykinin release (Busse & Lamontagne, 1991). Mediators such as ATP and substance P potentially provide a countercurrent mechanism which couples metabolism and perfusion as they are also released by hypoxia, and EDRF released by venules can diffuse and influence adjacent arterioles (Falcone & Bohlen, 1990; Ralevic et al., 1992).

EDRF release stimulated by flow exerts an inhibitory prejunctional effect on catecholamine release from adrenergic nerves, thereby contributing to a negative feedback loop which potentially stabilises the interaction between intravascular flow and sympathetic vasoconstriction (Fig. 2) (Tesfamarium & Cohen, 1988). There is also evidence that acetylcholine released from adventitial nerves diffuses to the endothelium in concentrations which stimulate EDRF release in the coronary and pulmonary circulations (Broten et al., 1992; McMahon et al., 1992). A similar mechanism has been reported in respect of Substance P in skeletal muscle (Persson et al., 1991). Additionally, NO released from non-cholinergic non-adrenergic (NANC) nerves inhibits contraction in various artery types (Toda & Okamura, 1990; Ahlner et al., 1991). It is presently unknown whether the mechanical forces which act on the vessel wall modulate these neurogenic mechanisms (Fig. 2).

The influence of time-averaged shear stress on EDRF release can be demonstrated by addition of dextran to the perfusate passing through an endothelium-intact donor vessel and monitoring the subsequent relaxation of an endothelium-denuded detector vessel segment located 'downstream' in cascade bioassay. Such relaxation is particularly evident when flow is made pulsatile, being maximal at pulse frequencies between 4 and 6 Hz in conduit arteries from the rat, rabbit and pig (Fig. 3) (Hutcheson & Griffith, 1991). This range of frequencies is generally

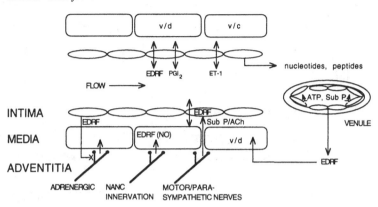

Fig. 2. There are complex interactions between blood flow and the three layers of the arterial wall. Shear stress induces release of the vasodilators (v/d) EDRF and prostacyclin (PGI₂), and the vasoconstrictor (v/c) peptide endothelin (ET-1). EDRF inhibits catecholamine release from sympathetic constrictor nerves, non-adrenergic non-cholinergic (NANC) nerves release NO, and neurotransmitters such as Substance P and acetylcholine can stimulate the endothelium to synthesise EDRF. Nucleotides and peptides may cause EDRF release from venules which potentially influence perfusion by diffusing to adjacent arterioles. The endothelium and smooth muscle of the media are not coupled electrically so that their interaction depends upon diffusible mediators.

above the resting heart rate in these species, suggesting that EDRF release evoked by tachycardia could contribute to exercise-related vasodilatation. In the coronary circulation EDRF release is also enhanced by the periodic compression imposed by myocardial contraction (Lamontagne, Pohl & Busse, 1992).

A flow-dependent response mediated by EDRF has also been demonstrated in small arteries and arterioles, potentially coupling changes in resistance in different parts of the vascular bed to changes in flow (Griffith *et al.*, 1987b, 1989; Kuo, Davis & Chilian, 1990; Pohl *et al.*, 1991). Many vascular beds autoregulate flow and are thus able to maintain constant perfusion at different input pressures. This phenomenon is mediated largely through feedback from vasodilator metabolites and by myogenic constrictor and dilator responses to rises and decreases in intravascular pressure, respectively. Flow-dependent dilatation will influence both mechanisms: a fall in distal resistance will reduce the resistance of more proximal 'feed' vessels through EDRF release stimulated by the associated rise in flow, and autoregulation is enhanced the augmentation

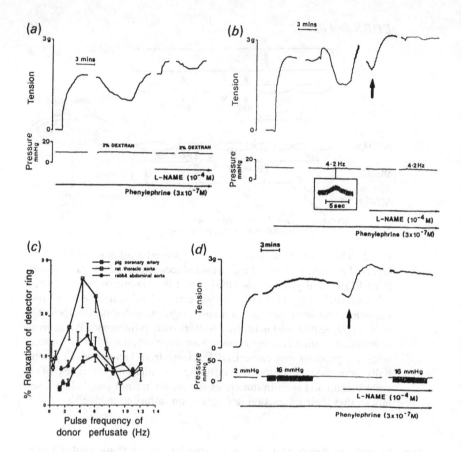

Fig. 3. Cascade bioassay experiments illustrating the influence of flow on EDRF release from isolated endothelium-intact segments of donor artery (at a constant mean flow of 9 ml min^{-1}). The detector tissue was an endothelium-denuded rabbit aortic ring preconstricted by phenylephrine. (a) Addition of dextran 80 to the perfusate caused reversible relaxation of the detector tissue which is attributable to an increase in viscosity and was attenuated by N^G-nitro-L-arginine methyl ester (L-NAME), an inhibitor of NO synthase. (b) Increasing the pulse frequency of the flow from 0.1 to 4.2 Hz at a constant pulse pressure amplitude of c. 2 mm Hg caused reversible relaxation of the detector that was abolished by L-NAME. The initial rise in tension caused by L-NAME (arrow) results from blockade of basal EDRF release. (c) Percentage relaxation of detector as a function of pulse frequency for three artery types from pig, rat and rabbit. In each case EDRF release was maximal over the range 4–6 Hz. (d) Increasing the amplitude of the pressure pulse from 2 to 16 mm Hg enhanced detector tone, an effect attributable to diminished EDRF release and thus abolished by L-NAME. Again inhibition of basal EDRF activity caused an initial increase in tension (arrow).

of myogenic responses following inhibition of EDRF activity (Griffith & Edwards, 1990a). Flow-related EDRF activity also serves to reduce cardiac work relative to perfusion, both by limiting the rise in pressure which would otherwise accompany increased flow and by maintaining an 'optimal' geometrical relationship between the diameters of the parent and daughter vessels at bifurcation points, which minimises hydraulic power losses (Griffith *et al.*, 1987b; Griffith & Edwards, 1990b). It may also maintain constancy in the distribution of microvascular flow at different flow rates by preventing 'steal' (Griffith *et al.*, 1987b). Loss of EDRF activity thus induces mismatches between metabolism and perfusion in skeletal muscle by reducing oxygen uptake and widening heterogeneity of tissue pO_2, and impairs pulmonary ventilation/perfusion matching (Pohl & Busse, 1988; Wiklund *et al.*, 1990).

EDRF may also participate in the alterations in vascular responsiveness and structure which follow long-term changes in flow and are thought to involve normalisation of endothelial shear stress (Kamiya & Togawa, 1980). Indeed, expression of endothelial cNOS mRNA is stimulated by increased fluid shear stress (Nishida *et al.*, 1992), thus explaining how basal EDRF release can be physiologically 'up-regulated' by creation of an arterio-venous fistula (Miller & Vanhoutte, 1988). The situation with agonist-induced release is more complex, however, as endothelium-dependent relaxation to acetylcholine and α_2 adrenergic receptor stimulation is enhanced by chronically increased flow, whereas that to bradykinin and ATP is not (Miller & Burnett, 1992). The mechanisms responsible for the adaptive decreases in diameter which follow chronically reduced flow are also endothelium-dependent, and may be associated with a reduction in DNA levels endothelial cell numbers, elastin content and wall mass (Langille & O'Donnell, 1986; Langille, Bendeck & Keeley, 1989). This could again involve EDRF as exogeneous NO reportedly inhibits angiogenesis and proliferation of cultured vascular smooth muscle (Garg & Hassid, 1989; Southgate & Newby, 1991; Pipili-Synetos *et al.*, 1993). Shear stress also modulates endothelial expression of platelet-derived growth factor mRNA and transforming growth factor-β1 mRNA, which affect both vascular cell growth and extracellular matrix production (Hsieh *et al.*, 1991; Ohno *et al.*, 1992b), so that vascular remodelling to altered flow is likely to be multifactorial in aetiology.

EDRF and pressure

Sustained increases in transmural pressure attenuate basal and acetylcholine-evoked EDRF release from intact arteries, and basal NO release from endothelium cultured on a flat glass surface increases at low and decreases at high pressures, being constant over the range 80–120 mm Hg

(Rubanyi, 1988; Hishikawa *et al.*, 1992). These observations could explain the blunting of flow-dependent dilatation observed at high transmural pressures in the femoral artery of the dog *in vivo* (Hilton, 1959). EDRF release is also inversely proportional to the amplitude of the pressure pulse in preparations perfused with pulsatile flow at constant mean flow and pulse frequency (Fig. 3). This may result, at least in part, from a fall in endothelial shear stress associated with an endothelium-independent increase in vessel calibre which reflects an intrinsic smooth muscle response to pulsatile forces (Hutcheson & Griffith, 1991). In a compliant hydrodynamically damped cardiovascular system at constant cardiac output *in vivo*, both stroke volume and the amplitude of pressure oscillations will be inversely proportional to heart rate, so that the frequency-dependent effects of flow and pressure on EDRF activity may be synergistic. Further evaluation of these phenomena is necessary, however, as pressure does not appear to influence basally released EDRF from endothelial cells cultured on microcarrier beads or that evoked by the calcium ionophore A23187 (Rubanyi *et al.*, 1991; Kelm *et al.*, 1991).

K^+ channels and EDRF release

Pharmacological evaluation of the contribution of hyperpolarising K^+ channels suggests that the Ca^{2+}-activated (K_{Ca}), but not the ATP-dependent (K_{ATP}) subtype is involved in EDRF release stimulated by the pulsatile component of flow, whereas both may be involved in the response to time-averaged shear stress. Thus, frequency-dependent EDRF release from rabbit aorta is antagonised by apamin, a selective blocker of low-conductance K_{Ca} channels and charybdotoxin (ChTX), which blocks medium- and high-conductance K_{Ca} channels, but not by glibenclamide, which selectively blocks K_{ATP} channels (Fig. 4). In contrast EDRF release evoked from rabbit aorta by increased perfusate viscosity and by mechanically stirring cultured bovine aortic endothelial cells is attenuated by ChTX, but not by apamin (Fig. 5; Cooke *et al.*, 1991). Differences in response to oscillatory and time-averaged shear forces are suggested by attenuated viscosity-related EDRF release in the

---

Fig. 4. Effects of (*a*) apamin (100 nM), (*b*) charybdotoxin (ChTX, 1 nM) and (*c*) glibenclamide (10 μM) on relaxation of the detector tissue to increased pulse frequency of flow in cascade bioassay (*n* = 5 in each case). Glibenclamide failed to attenuate the frequency-dependent effect, which thus involves K_{Ca} but not K_{ATP} channels (* denotes significantly different from control, $p < 0.05$).

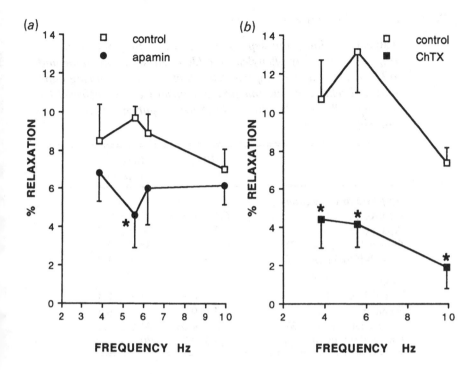

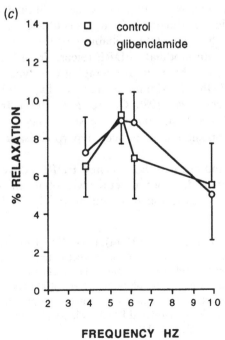

Table 1. *Endothelium-dependent relaxations of rabbit aortic rings constricted by phenylephrine (300 nM) to acetylcholine and ATP (expressed as − log [concentration of drug] inducing half maximal relaxation and maximal response) were not affected by apamin (100 nM), charybdotoxin (1 nM) or glibenclamide (10 μM).*

	EC_{50}	Maximal relaxation
Response to acetylcholine		
Control ($n = 13$)	6.4 ± 0.1	79.5 ± 3.8
Apamin ($n = 3$)	6.3 ± 0.03	66.0 ± 5.7
Glibenclamide ($n = 3$)	6.7 ± 0.1	84.2 ± 7.9
Charybdotoxin ($n = 3$)	6.5 ± 0.1	69.5 ± 1.9
Response to ATP		
Control ($n = 12$)	4.5 ± 0.3	35.3 ± 2.4
Apamin ($n = 3$)	4.7 ± 0.2	40.4 ± 9.7
Glibenclamide ($n = 3$)	4.3 ± 0.03	22.6 ± 3.1
Charybdotoxin ($n = 3$)	4.6 ± 0.3	30.4 ± 4.3

presence of glibenclamide, which implicates the additional involvement of a K_{ATP} channel (Fig. 5). Relatively high levels of shear stress may be necessary to increase the open state probability of this channel subtype, as glibenclamide does not modulate EDRF release from cultured endothelial cells exposed to a shear stress averaging 0.3 dyne cm^{-2}, which produces c. 10% of the maximum achievable hyperpolarisation in response to flow (Olesen *et al.*, 1988; Cooke *et al.*, 1991). In the experiments of Fig. 5 the highest calculated shear stress was 4–9 dyne cm^{-2}, which results in near maximal flow-induced hyperpolarisation (Olesen *et al.*, 1988).

In contrast to the findings with flow-related EDRF activity, blockade of K_{Ca} and K_{ATP} channels does not affect acetylcholine- or ATP-evoked endothelium-dependent relaxation in rabbit aorta (Table 1) or EDRF

Fig. 5. Effects of (*a*) apamin (100 nM), (*b*) ChTX (1 nM) and (*c*) glibenclamide (10 μM) on relaxation of the detector tissue as a function of perfusate viscosity which was manipulated by inclusion of 1–4% dextran 80 ($n = 5$ in each case). The inhibitory effect of glibenclamide but not of apamin implies the involvement of distinct subpopulations of K_{Ca} and K_{ATP} channels in viscosity-related EDRF release (* denotes significantly different from control, $p < 0.05$).

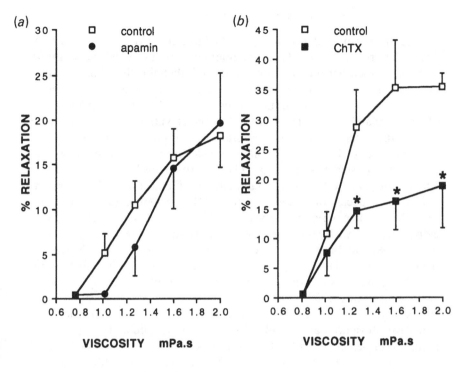

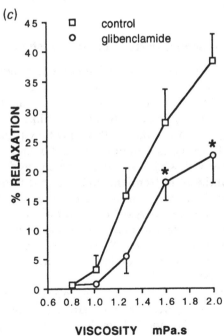

release from cultured porcine endothelial cells (Cooke *et al.*, 1991). These differences may reflect the relative importance of the multiple pathways which permit Ca^{2+} entry into endothelial cells. Thus, agonists may stimulate Ca^{2+} influx through K^+ channel independent mechanisms such as receptor-operated influx (Johns *et al.*, 1987), and replenishment of endothelial stores depleted of Ca^{2+} by the elevation of IP_3 levels which follows the receptor–G protein interaction and phosphoinositide hydrolysis, may either be direct or follow translocation from an intermediate pool to the IP_3-sensitive store via a GTP-sensitive mechanism (Schilling & Elliot, 1992). Indeed, there is accumulating evidence for significant mechanistic and species differences in the EDRF response to agonists and flow. In rabbit ear resistance arteries inhibition of flow-related EDRF activity by L-NAME (nitro-arginine methyl ester) is reversed by L-arginine, whereas inhibition of the response to acetylcholine by L-NAME is irreversible (Randall & Griffith, 1991). This contrasts with cat femoral artery in which flow-dependent dilatation is not attenuated by L-NAME at concentrations which reversibly inhibit the response to acetylcholine (Melkumyants *et al.*, 1992). Furthermore, inhibition of mitochondrial oxidative phosphorylation and depletion of intracellular Ca^{2+} stores by thapsigargin selectively inhibit agonist-evoked but not basal/flow-induced EDRF release (Griffith, Edwards & Henderson, 1987a; MacArthur *et al.*, 1993).

Other endothelium-derived dilator substances

Evidence that prostanoids may also contribute to flow-dependent dilatation has been obtained in the rat skeletal microcirculation *in vivo* (Koller & Kalley, 1990). The major vasodilator prostanoid, prostacyclin (PGI_2), derives mainly from the endothelium, and like EDRF, its release is greater with pulsatile rather than with steady flow (Pohl *et al.*, 1986). Shear-induced PGI_2 synthesis requires Ca^{2+} influx and is characterised by sustained release following an initial burst in production, whereas agonist-induced synthesis is transient and dependent on elevation of $[Ca^{2+}]_i$ through mobilisation of intracellular stores (Bhagyalakshmi & Frangos, 1989). PGI_2 elevates cAMP levels within cells and may therefore potentially contribute to phosphorylation and inhibition of the endothelial cNOS (Nishida *et al.*, 1992). Shear forces also evoke release of the potent constrictor peptide endothelin-1 from endothelial cells (Yoshizumi *et al.*, 1989).

Several groups have postulated the existence of a further diffusible vasodilator, endothelium-derived hyperpolarising factor, EDHF, on the basis that in some vessels endothelium-dependent hyperpolarisation of vascular smooth muscle is reduced but not abolished by inhibition of EDRF activity, and there is no electrical coupling between the two cell types (Tare *et al.*, 1990; Garland & McPherson, 1992; Segal & Bény, 1992). Indeed, acetylcholine may stimulate EDHF release via M_1 muscarinic receptors and EDRF activity via M_2 receptors (Komori & Suzuki, 1987). Regional and species heterogeneity in the participating mechanisms is evident in that non-EDRF mediated endothelium-dependent hyperpolarisation may involve K_{ATP} or K_{Ca} channels in different artery types (Bray & Quast, 1991; Brayden, 1991; Chen & Cheung, 1992). While it has been speculated that the biphasic vasodilator response to acetylcholine *in vivo* consists of an initial EDHF-mediated transient followed by a more prolonged EDRF-mediated phase (Aisaka *et al.*, 1989), a specific role for EDHF in the response to flow has yet to be demonstrated.

Conclusions

The realisation that the endothelium exerts a pivotal role in the control of vasomotor tone has focused attention on the intimate communication between the different cell types present in the arterial wall. From a chemical point of view NO may be one of the simplest known biological mediators, but its physiological consequences are complex and have added a previously unsuspected dimension to the mechanisms which mediate the interaction between flow and tissue perfusion.

References

Ahlner, J., Ljusegren, M. E., Grundstrom, N. & Axelsson, K. L. (1991). Role of nitric oxide and cyclic GMP as mediators of endothelium-independent neurogenic relaxation in bovine mesenteric artery. *Circulation Research* **68**, 756–62.

Aisaka, K., Gross, S. S., Griffith, O. W. & Levi, R. (1989) L-arginine availability determines the duration of acetylcholine-induced systemic vasodilatation *in vivo*. *Biochemical and Biophysical Research Communications* **163**, 710–17.

Assender, J. W., Southgate, K. M. & Newby, A. C. (1991). Does nitric oxide inhibit smooth muscle proliferation? *Journal of Cardiovascular Pharmacology* **17** (Suppl. 3), S104–7.

Bhagyalakshmi, A. & Frangos, J. A. (1989). Mechanism of shear-induced prostacyclin production in endothelial cells. *Biochemical and Biophysical Research Communications* **158**, 31–7.

Blayney, L. M., Gapper, P. W. & Newby, A. C. (1991). Inhibition of a receptor-operated calcium channel in pig aortic microsomes by a cyclic GMP-dependent protein kinase. *Biochemical Journal* **273**, 803–6.

74   T. M. GRIFFITH & I. R. HUTCHESON

Bodin, P., Bailey, D. & Burnstock, G. (1991). Increased flow-induced ATP release from isolated vascular endothelial cells but not smooth muscle cells. *British Journal of Pharmacology* **103**, 1203–5.

Bray, K. & Quast, U. (1991). Differences in the K^+-channels opened by cromakalim, acetylcholine and substance P in rat aorta and porcine coronary artery. *British Journal of Pharmacology* **102**, 585–94.

Brayden, J. E. (1991). Hyperpolarization and relaxation of resistance arteries in response to adenosine diphosphate. *Circulation Research* **69**, 1415–20.

Broten, T. P., Miyashiro, J. K., Moncada, S. & Feigel, E. O. (1992). Role of endothelium-derived relaxing factor in parasympathetic coronary vasodilation. *American Journal of Physiology* **262**, H1579–84.

Busse, R. & Lamontagne, D. (1991). Endothelium-derived bradykinin is responsible for the increase in calcium produced by angiotensin-converting enzyme inhibitors in human endothelial cells. *Naunyn-Schmiedeberg's Archives of Pharmacology* **344**, 126–9.

Chen, G. & Cheung, D. W. (1992). Characterization of acetylcholine-induced membrane hyperpolarization in endothelial cells. *Circulation Research* **70**, 257–63.

Collins, P., Griffith, T. M., Henderson, A. H. & Lewis, M. J. (1986). Endothelium-derived relaxing factor alters calcium fluxes in rabbit aorta: a cyclic guanosine monophosphate-mediated effect. *Journal of Physiology (London)* **381**, 427–37.

Cooke, J. P., Rossitch, E., Andon N. A., Loscalzo, J. & Dzau, V. J. (1991). Flow activates an endothelial potassium channel to release an endogeneous vasodilator. *Journal of Clinical Investigation* **88**, 1663–71.

Falcone, J. C. & Bohlen, H. G. (1990). EDRF from fat intestine and skeletal muscle venules causes dilation of arterioles. *American Journal of Physiology* **258**, H1515–23.

Fiscus, R. R., Rapoport, R. M. & Murad, F. (1984). Endothelium-dependent and nitrovasodilator-induced activation of cyclic GMP-dependent protein kinase in rat aorta. *Journal of Cyclic Nucleotide Protein Phosphorylation Research* **9**, 415–25.

Furchgott, R. F. & Zawadski, J. V. (1980). The obligatory role of endothelial cells in the relaxation of arterial smooth muscle by acetylcholine. *Nature* **288**, 373–6.

Garg, U. C. & Hassid, A. (1989). Nitric oxide-generating vasodilators and 8-bromo-cyclic guanosine monophosphate inhibit mitogenesis and proliferation of cultured rat vascular smooth muscle cells. *Journal of Clinical Investigation* **83**, 1774–7.

Garland, C. J. & McPherson, G. A. (1992). Evidence that nitric oxide does not mediate the hyperpolarisation and relaxation to acetylcholine in the rat. *British Journal of Pharmacology* **105**, 429–35.

Geiger, R. V., Berk, B. C., Alexander, R. W. & Nerem, R. W. (1992). Flow-induced calcium transients in single endothelial cells: spatial and temporal analysis. *American Journal of Physiology* **262**, C1411–17.

Griffith, T. M. & Edwards, D. H. (1990a). Myogenic autoregulation of flow may be inversely related to EDRF activity. *American Journal of Physiology* **258**, H1171–80.

Griffith, T. M. & Edwards, D. H. (1990b). Basal EDRF activity helps to keep the geometrical configuration of arterial bifurcations close to the Murray optimum. *Journal of Theoretical Biology* **146**, 545–73.

Griffith, T. M., Edwards, D. H., Davies, R. Ll., Harrison, T. J. & Evans, K. T. (1987b). EDRF coordinates the behaviour of vascular resistance vessels. *Nature* **329** 442–5.

Griffith, T. M., Edwards, D. H., Davies, R. Ll., Henderson, A. H. (1989). The role of EDRF in flow distribution: a microangiographic study of the rabbit isolated ear. *Microvascular Research* **37**, 162–77.

Griffith, T. M., Edwards, D. H. & Henderson, A. H. (1987a). Unstimulated production of endothelium-derived relaxing factor is independent of mitochondrial ATP generation. *Cardiovascular Research* **21**, 565–8.

Griffith, T. M., Edwards, D. H., Lewis, M. J., Newby, A. C. & Henderson, A. H. (1984). The nature of endothelium-derived vascular relaxant factor. *Nature* **308**, 645–7.

Griffith, T. M., Edwards, D. H., Newby, A. C., Lewis, M. J. & Henderson, A. H. (1986). Production of endothelium-derived relaxant factor is dependent on oxidative phosphorylation and extracellular calcium. *Cardiovascular Research* **20**, 7–12.

Hilton, S. M. (1959). A peripheral arterial conducting mechanism underlying dilatation of the femoral artery and concerned in functional vasodilatation in skeletal muscle. *Journal of Physiology (London)* **149**, 93–111.

Hirata, M., Kohse, K. P., Chang, C.-H., Ikebe, T. & Murad, F. (1990). Mechanism of cyclic GMP inhibition of inositol phosphate formation in rat aortic segments and cultured bovine aortic smooth muscle cells. *Journal of Biological Chemistry* **265**, 1268–73.

Hishikawa, K., Nakaki, T., Suzuki, H., Saruta, T. & Kato, R. (1992). Pure transmural pressure inhibits nitric oxide from cultured human endothelial cells. *Journal of Vascular Research* **29**, 36.

Holtz, J., Forstermann, U., Pohl, U., Giesler, M. & Bassenge, E. (1984). Flow-dependent endothelium-mediated dilatation of epicardial coronary arteries in conscious dogs: effect of cyclooxygenase inhibition. *Journal of Cardiovascular Pharmacology* **6**, 1161–9.

Hsieh, H.-J., Li, N.-Q., & Frangos, J. A. (1991). Shear stress increases endothelial platelet-derived growth factor mRNA levels. *American Journal of Physiology* **260**, H642–6.

Hutcheson, I. R. & Griffith, T. M. (1991). Release of endothelium-derived relaxing factor is modulated both by frequency and amplitude of pulsatile flow. *American Journal of Physiology* **261**, H257–62.

Itoh, T., Seki, N., Suzuki, S., Ito, S., Kajikuri, J. & Kuriyama, H.

(1992). Membrane hyperpolarisation inhibits agonist-induced synthesis of inositol 1,4,5-trisphosphate in rabbit mesenteric artery. *Journal of Physiology* **451**, 307–28.

Johns, A., Lategan, T. W., Lodge N. J., Ryan, U. S., Van Breemen, C. & Adams, D. J. (1987). Calcium entry through receptor-operated channels in bovine artery endothelial cells. *Tissue and Cell* **19**, 733–45.

Joulou-Schaeffer, G., Gray, G. A., Fleming, I., Schott, C., Parratt, J. R. & Stocklet, J.-C. (1990). Loss of vascular responsiveness induced by endotoxin involves the L-arginine pathway. *American Journal of Physiology* **259**, H1038–43.

Kamiya, A. & Togawa, T. (1980). Adaptive regulation of wall shear stress to flow change in the canine carotid artery. *American Journal of Physiology* **239**, H14–21.

Kelm, M., Feelisch, M., Deussen, A., Strauer, B. E. & Schrader, J. (1991). Release of endothelium derived nitric oxide in relation to pressure and flow. *Cardiovascular Research* **25**, 831–6.

Koller, A. & Kaley, G. (1990). Prostaglandins mediate arteriolar dilation to increased blood flow velocity in skeletal muscle microcirculation. *Circulation Research* **67**, 529–34.

Komori, K. & Suzuki, H. (1987). Heterogeneous distribution of muscarinic receptors in the rabbit saphenous artery. *British Journal of Pharmacology* **92**, 657–64.

Kuo, L., Davis, M. J. & Chilian, W. M. (1990). Endothelium-dependent, flow-induced dilation of isolated coronary arterioles. *American Journal of Physiology* **259**, H1063–70.

Lamontagne, D., Pohl, U. & Busse, R. (1992). Mechanical deformation of vessel wall and shear stress determine the basal release of endothelium-derived relaxing factor in the intact rabbit coronary vascular bed. *Circulation Research* **70**, 123–30.

Langille, B. L., Bendeck, M. P. & Keeley, F. W. (1989). Adaptations of carotid arteries of young and mature rabbits to reduce carotid blood flow. *American Journal of Physiology* **256**, H931–9.

Langille, B. L. & O'Donnell, F. (1986). Reductions in arterial diameter produced by chronic decreases in blood flow are endothelium-dependent. *Science* **231**, 405–7.

Lansman, J. B., Hallam, T. J. & Rink, T. J. (1987). Single stretch-activated ion channels in vascular endothelial cells as mechano-transducers? *Nature* **325**, 811–13.

Lückhoff, A. & Busse, R. (1990). Calcium influx into endothelial cells and formation of endothelium-derived relaxing factor is controlled by the membrane potential. *Pflügers Archiv* **416**, 305–11.

Lückhoff, A. & Clapham, D. E. (1992). Inositol 1,3,4,5-tetrakisphosphate activates an endothelial Ca^{2+}-permeable channel. *Nature* **355** 356–8.

MacArthur, H., Hecker, M., Busse, R. & Vane, J. R. (1993). Selective inhibition of agonist induced but not shear stress-dependent release of endothelial autocoids by thapsigargin. *British Journal of Pharmacology* **108**, 100–5.

McMahon, T. J., Hood, J. S. & Kadowitz, P. J. (1992). Pulmonary vasodilator response to vagal stimulation is blocked by N^W-Nitro-L-arginine methyl ester in the cat. *Circulation Research* **70**, 364–9.

Malinski, T. & Taha, Z. (1992). Nitric oxide release from a single cell measured *in situ* by a porphyrin based microsensor. *Nature* **358**, 676–8.

Melkumyants, A. M. & Balashov, S. A. (1990). Effect of blood viscosity on arterial flow induced dilator response. *Cardiovascular Research* **24**, 165–8.

Melkumyants, A. M., Balashov, S. A., Klimachev, A. N., Kartamyshev, S. P. & Khayutin, V. M. (1992). Nitric oxide does not mediate flow induced endothelium-dependent arterial dilatation in the cat. *Cardiovascular Research* **26**, 156–60.

Miller, V. M. & Burnett, J. C. (1992). Modulation of NO and endothelin by chronic increases in blood flow in canine femoral arteries. *American Journal of Physiology* **263**, H103–8.

Miller, V. M. & Vanhoutte, P. M. (1988). Enhanced release of endothelium-derived factor(s) by chronic increases in blood flow. *American Journal of Physiology* **255**, H446–51.

Moncada, S., Palmer, R. M. J. & Higgs, E. A. (1991). Nitric oxide physiology, pathophysiology, and pharmacology. *Pharmacology Review* **43**, 109–42.

Nakache, M. & Gaub, H. E. (1988). Hydrodynamic hyperpolarisation of endothelial cells. *Proceedings of the National Academy of Sciences USA* **85**, 1841–3.

Nishida, K., Harrison, D. G., Navas, J. P., Fisher A. A., Dockery, S. P., Uematsu, M., Nerem, R. M., Alexander, R. W. & Murphy, T. J. (1992). Molecular cloning and characterization of the constitutive bovine aortic endothelial cell nitric oxide synthase. *Journal of Clinical Investigation* **90**, 2092–6.

Nishikawa, M., DeLanerolle, P., Lincoln, T. H. M. & Adelstein, R. S. (1984). Phosphorylation of mammalian myosin light chain kinases by the catalytic subunit of cyclic AMP-dependent protein kinase and by cyclic GMP-dependent protein kinase. *Journal of Biological Chemistry* **259**, 8429–36.

Nishimura, J. & Van Breemen, C. (1989). Direct regulation of smooth muscle contractile elements by second messengers. *Biochemical and Biophysical Research Communications* **163**, 929–35.

Nollert, M. U., Eskin, S. G. & McIntire, L. V. (1990). Shear stress increases inositol trisphosphate levels in human endothelial cells. *Biochemical and Biophysical Research Communications* **170**, 281–7.

Ohno, M., Gibbons, G. H., Cooke, J. P. & Dzau, V. J. (1992a). Endo-

78   T. M. GRIFFITH & I. R. HUTCHESON

thelial potassium channel sensitive to pertussis toxin mediates the release of EDRF by shear stress. *Journal of Vascular Research* **29**, 174.

Ohno, M., Gibbons, G. H., Lopez, F., Cooke, J. P. & Dzau, V. J. (1992b). Shear stress induces transforming growth factor-β1 (TGF-β1) expression via a flow-activated potassium channel. *Journal of Vascular Research* **29**, 175.

Olesen, S.-P., Clapham, D. E. & Davies, P. F. (1988). Haemodynamic shear stress activates a K^+ current in vascular endothelial cells. *Nature* **331**, 168–70.

Persson, M. G., Hedqvist, P. & Gustafsson, L. E. (1991). Nerve-induced tachykinin-mediated vasodilatation in skeletal muscle is dependent on nitric oxide formation. *European Journal of Pharmacology* **205**, 295–301.

Pipili-Synetos, E., Sakkoula, E. & Maragoudakis, M. E. (1993). Nitric oxide is involved in the regulation of angiogenesis. *British Journal of Pharmacology* **108**, 855–7.

Pohl, U. & Busse, R. (1988). Reduced nutritional blood flow in auto-perfused rabbit hindlimbs following inhibition of endothelial vasomotor function. In *Resistance Arteries*, ed. W. Halpern *et al.*, pp. 10–16. Ithaca, NY: Perinatology Press.

Pohl, U., Herlan, K., Huang, A. & Bassenge, E. (1986). Pulsatile perfusion stimulates the release of endothelial autoacoids. *Journal of Applied Cardiology* **1**, 215–35.

Pohl, U., Herlan, K., Huang, A. & Bassenge, E. (1991). EDRF-mediated, shear-induced dilation opposes myogenic vasoconstriction in small rabbit arteries. *American Journal of Physiology* **261**, H2016–23.

Popescu, L. M., Panoiu, C., Hinescu, M. & Nutu, O. (1985). The mechanism of cGMP-induced relaxation in vascular smooth muscle. *European Journal of Pharmacology* **107**, 393–4.

Ralevic, V., Lincoln, J. & Burnstock, G. (1992). Release of vasoactive substances from endothelial cells. In *Endothelial Regulation of Vascular Tone*, ed. U. S. Ryan & G. M. Rubanyi, pp. 297–328. New York: Marcel Dekker.

Ralevic, V., Milner, P., Hudlicka, O., Kristek, F. & Burnstock, G. (1990). Substance P is released from the endothelium of normal and capsaicin-treated rat hind-limb vasculature, *in vivo*, by increased flow. *Circulation Research* **66**, 1178–83.

Randall, M. D. & Griffith, T. M. (1991). Differential effects of L-arginine on the inhibition of N^G-nitro-L-arginine methyl ester of basal and agonist-stimulated EDRF activity. *British Journal of Pharmacology* **104**, 743–9.

Rapoport, R. M. (1986). Cyclic guanosine monophosphate inhibition of contraction may be mediated through inhibition of phosphatidylinositol hydrolysis in rat aorta. *Circulation Research* **58**, 407–10.

Rosales, O. R. & Sumpio, B. E. (1992). Changes in cyclic strain increase inositol trisphosphate and diacylglycerol in endothelial cells. *American Journal of Physiology* **262**, C956–62.

Rubanyi, G. M. (1988). Endothelium-dependent pressure-induced contraction of isolated canine carotid arteries. *American Journal of Physiology* **255**, H783–8.

Rubanyi, G. M., Freay, A. D. Johns, A. & Van Breemen, C. (1991). Elevated transmural pressure inhibits the release of EDRF by mechanisms similar to high K$^+$ and barriers. In *Resistance Arteries, Structure and Function*, ed. M. J. Mulvaney, C. Aalkjaer, A. M. Heagerty, N. C. B. Nyborg & S. Strandgaards, pp. 226–32. Amsterdam: Excerpta Medica.

Rubanyi, G. M., Romero, J. C. & Vanhoutte, P. M. (1986). Flow-induced release of endothelium-derived relaxing factor. *American Journal of Physiology* **250**, H1145–9.

Schilling, W. P. & Elliot, S. J. (1992). Ca^{2+} signalling mechanisms of vascular endothelial cells and their role in oxidant-induced endothelial cell dysfunction. *American Journal of Physiology* **262**, H1617–30.

Schretzenmayr, A. (1933). Über kreislaufregulatorische Vorgänge an den grossen Arterien bei der Muskelarbeit. *Pflügers Archiv* **232**, 743–8.

Schwartz, G., Callewaert, G., Droogmans, C. & Nilius, B. (1992). Shear stress-induced calcium transients in endothelial cells from human umbilical cord veins. *Journal of Physiology* **458**, 527–38.

Segal, S. S. & Beny, J.-L. (1992). Intracellular recording and dye transfer in arterioles during blood flow control. *American Journal of Physiology* **263**, H1–7.

Sinoway, L. I., Hendrickson, C., Davidson, W. R., Jr, Prophet, S. & Zelis, R. (1989). Characteristics of flow-mediated brachial artery vasodilatation in human subjects. *Circulation Research* **64**, 32–42.

Tare, M., Parkington, H. C., Coleman, H. A., Neild, T. O. & Dusting, G. J. (1990). Hyperpolarization and relaxation of arterial smooth muscle caused by nitric oxide derived from the endothelium. *Nature* **346**, 69–71.

Tesfamarium, B. & Cohen, R. A. (1988). Inhibition of adrenergic vasoconstriction by endothelial cell shear stress. *Circulation Research* **63**, 720–5.

Toda, N. & Okamura, T. (1990). Possible role of nitric oxide in transmitting information from vasodilator nerve to cerebroarterial muscle. *Biochemical and Biophysical Research Communications* **170**, 308–13.

Twort, C. H. C. & Van Breemen, C. (1988). Cyclic guanosine monophosphate-enhanced sequestration of Ca^{2+} by sarcoplasmic reticulum in vascular smooth muscle. *Circulation Research* **62**, 961–4.

Wiklund, N. P., Persson, M. G., Gustafsson, L. E., Moncada, S. & Hedqvist, P. (1990). Modulatory role of endogenous nitric oxide in pulmonary circulation *in vivo*. *European Journal of Pharmacology* **185**, 123–4.

Xie, Q.-W., Cho, H. J., Calaycay, J., Mumford, R. A., Swiderek, K. M., Lee, T. D., Ding, A., Troso, T. & Natham, C. (1992). Cloning and characterization of inducible nitric oxide synthase from mouse macrophages. *Science* **256**, 225–8.

Yoshizumi, M., Kurihara, H., Sugiyama, T., Takaku, F., Yanagisawa, M., Masaki, T. & Yazaki, Y. (1989). Hemodynamic shear stress stimulates endothelin production by cultured endothelial cells. *Biochemical and Biophysical Research Communications* **161**, 859–64.

G. GOLDSPINK, G.-F. GERLACH,
T. JAENICKE and P. BUTTERWORTH

Stretch, overload and gene expression in muscle

Skeletal muscle has considerable capacity to increase in mass, both during post-natal growth and even in the adult, to changed physical activity levels. For instance, during the rapid post-natal growth period, the male mouse is adding 25% new muscle per day (Goldspink & Griffin, 1973). In the adult rabbit, the tibialis anterior when subjected to stretch combined with electrical stimulation increases by 35% in mass within 4 days (Goldspink et al., 1992). This latter represents a synthetic rate of 30 000 myosin heavy chain (hc) molecules per nucleus per minute (Goldspink, 1985) which is remarkable but still only about one third of the synthetic capacity during the most rapid post-natal growth period.

As well as changes in muscle mass, changes in muscle phenotype often occur in response to physical signals (Gerlach et al., 1990; Goldspink et al., 1992). Mammalian skeletal muscles consist of populations of slow-contracting, oxidative fibres which are adapted for slow repetitive or postural type contractile activity, and fast-contracting fibres that are recruited for fast phasic movements. The muscle fibre types differ phenotypically in that they express different subsets of myofibrillar isoform genes as well as different types and levels of metabolic enzymes. The inherent ability of skeletal muscle to adapt to mechanical signals is related to its ability to switch on or switch off different isoform genes and to alter the general levels of expression of different subsets of genes. The fact that there are several myosin hc isoforms means that a muscle fibre can alter its contractile properties by rebuilding its myofibrils using a myosin hc with a slow or fast cross bridge cycling rate and changing its intrinsic velocity of contraction (V_{max}). We have focused our attention on the myosin hc genes because of the strong correlation between the V_{max} of the muscle fibres and their fast and slow myosin hc content (Reiser et al., 1985). The isoforms of myosin hc have been shown to be encoded

Society for Experimental Biology Seminar Series 54: *Biomechanics and Cells*, ed. F. Lyall & A. J. El Haj. © Cambridge University Press 1994, pp. 81–95.

by individual genes of a multigene family and their expression tightly regulated in a stage-specific and tissue-specific manner (Butler-Brown & Whalen, 1984; Mahdavi *et al.*, 1986; Weydert *et al.*, 1987) and influenced by hormones (Lompre, Nadal-Ginard & Mahdavi, 1984; Izumo, Nadal-Ginard & Mahdavi, 1986) and type of innervation (Pette & Vrbova, 1986). However, the influence of physical activity at the gene level was unclear. Therefore we have studied changes in gene expression of the fast and slow genes in response to the two main mechanical stimuli: stretch and force generation.

Stretch has been shown to be a very powerful stimulant of muscle growth and muscle protein synthesis. During post-natal growth, skeletal muscle fibres elongate by adding new sarcomeres serially to the ends of existing myofibrils (Griffin, Williams & Goldspink, 1971). Even mature muscles have been shown to be capable of adapting to a new functional length by adding or removing sarcomeres in series (Goldspink, 1984; Williams *et al.*, 1986) (Fig. 1). In this way sarcomere length is adjusted back to the optimum for force generation, velocity and hence power output. The stretch effect and the adaptation to an increased functional length is known to be associated with increased protein synthesis (Goldspink & Goldspink, 1986; Loughna, Goldspink & Goldspink, 1986) (Fig. 2). More recently, we have studied the way gene expression in muscle is influenced by stretch or the absence of stretch. This study commenced with a collaboration with Professor Nadal-Ginard's group at the Harvard Medical School, who were the first to produce the necessary gene probes to detect the specific mRNA of different myosin hc isoforms in rat skeletal muscle. Stretch or lack of stretch was achieved by casting the limb with the muscle in the shortened or lengthened position. Several interesting findings emerged from this study (Loughna *et al.*, 1986), including the fact that a slow soleus muscle, which does not normally express fast type IIb myosin hc genes, begins to transcribe the fast myosin hc gene after only a day if its muscle fibres are not stretched passively or are not producing force. It is interesting to note that under these conditions the muscle is actually undergoing atrophy even though initially there is abundant message available.

The converse situation was obtained when the fast tibialis anterior of the rabbit was subjected to passive stretch and to static force generation. In this case the slow genes were turned on and the fast myosin genes repressed. Also the tibialis anterior muscle underwent considerable rapid hypertrophy (35% in 4 days). Stretch combined with stimulation was achieved by using miniature stimulators and by immobilisation of the limb with the tibialis anterior in its lengthened position (Gregory, Low & Stirewalt, 1986). Several designs of stimulator circuit were used including

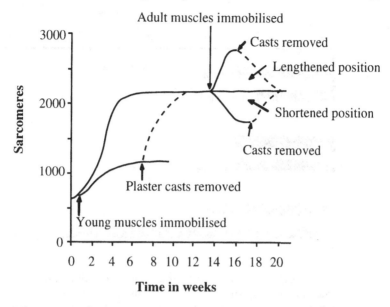

Fig. 1. A summary of the change in the number of sarcomeres in series along the length of single teased muscles fibres from the soleus muscle of mice of different ages. The post-natal increase is suppressed if the muscle in the young mouse is immobilised just after birth (using a plaster cast that is changed frequently so as not to interfere with bone growth). Under these circumstances, the tendons still grow and the muscle belly remains shorter than normal until the plaster cast is removed, when the fibres then produce sarcomeres in series at a very rapid rate until the sarcomere length is adjusted back to the optimum. Even in adult muscle the fibres can still adjust to a change in functional length by adding or removing sarcomeres in series. When the muscle is held in the stretched position by plaster cast it adds on 20–30% more sarcomeres in series within about a week. If the muscle is immobilised in its shortened position it takes off about 20–30% of sarcomeres. Both processes are reversible and the adult muscle soon adapts back to its normal functional length when the cast is removed. Data from Williams & Goldspink (1973).

some which gave low frequency continuous trains and some which gave higher frequency intermittent trains. 30 and 60 Hz intermittent circuits were designed to give the same number of pulses per minute as 2 and 10 Hz continuous stimulation circuits. In this way the hypothesis, that it is the number of pulses that is important, could be tested. In all cases the pulses were biphasic and the pulse voltage was adjustable up to a

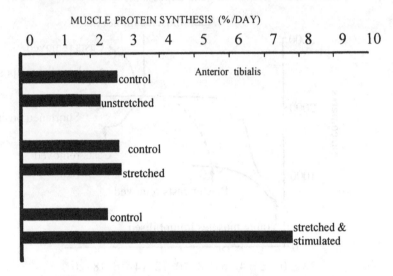

Fig. 2. Effect of stretch and stretch combined with stimulation on the rates of protein synthesis in the tibialis anterior muscle in the adult rabbit. Synthesis rates are given as per cent per day. Muscles immobilised in the unstretched position atrophy and, as discussed in the text, this is associated with a decreased rate of synthesis. Stretch on the other hand maintains the normal rate of synthesis. However, when stretch is combined with electrical stimulation (5 Hz continuous) the rate of protein synthesis is greatly increased (Goldspink & Goldspink, 1986). This correlates well with the finding that the mass of the tibialis anterior in the adult rabbit will increase by up to 35% within 4 days of being stretched and stimulated (Goldspink *et al.*, 1992).

maximum of 3 volts. To study the switches in gene expression, RNA was extracted from stretched, stimulated and contralateral control muscles. Northern blots (Fig. 3) were carried out under different stringency conditions (Goldspink *et al.*, 1992) using cDNA for the slow beta type myosin hc, originally cloned and characterised in the laboratory of Dr Zak (Umeda *et al.*, 1981), a cDNA specific for the fast type IIb myosin hc and also cDNA for myosin light chains 1 and 3 from Dr Whittinghoffer (Maeda, Sckziel & Whittinghoffer, 1987). As mentioned, stretch combined with electrical stimulation was also found to induce very rapid hypertrophy of the tibialis anterior in the adult animal. Both force generation and stretch are major factors in activating protein synthesis and the combination of these stimuli apparently has a pronounced synergistic effect. Associated with this very significant increase in muscle size there was a marked increase (up to 250%) in RNA content of the muscles

Stretch, overload and gene expression 85

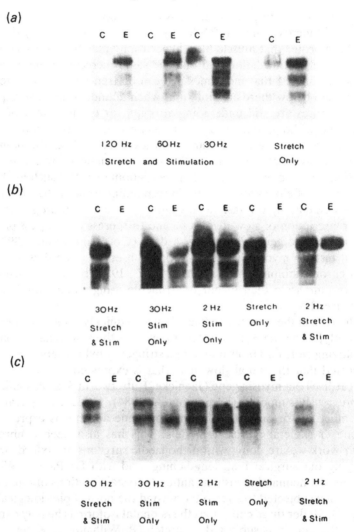

Fig. 3. Northern blots using RNA from stretched and stimulated rabbit tibialis anterior muscles after 4 days. (*a*) Beta cardiac general slow myosin gene probe showing activation of slow myosin hc. (*b*) Fast IIb myosin hc probe showing repression of this gene with stretch and/or stimulation. (*c*) Myosin light chain 1 & 3 probe showing repression of the myosin hc genes with stretch and/or stimulation (Goldspink *et al.*, 1992).

which was found to peak after 2 days of the commencement of stretch. This rapid increase in total RNA, which is presumably mainly ribosomal RNA, indicates that muscle fibre hypertrophy may be controlled mainly at the level of translation and that the rapid increase in the number of ribosomes means that more message can be translated into protein. As mentioned below there are situations when abundant message is present but the fibres are still undergoing atrophy, for example lack of stretch (with and without stimulation). This again indicates that fibre size may be regulated mainly at the translation level. As well as the number of ribosomes available to translate the available message, whatever it may be, there are factors implicated in translational control such as the concentrations of elongation factors. In responding to both stretch and electrical stimulation, the fast tibialis anterior becomes reprogrammed for the transcription of slow myosin hc and to repress the expression of the fast myosin hc gene within only 4 days (Goldspink *et al.*, 1992). The rabbit fast IIb myosin hc gene has been observed to be down-regulated with chronic stimulation (Brownson *et al.*, 1989) but on a considerably longer time scale than in our experiments using stimulation combined with stretch.

The reason the muscle becomes slower so quickly is more problematic. Recently, using the 3′ untranslated region of the slow beta cardiac gene employing Northern blots under high stringency hybridisation conditions, indicated that the initial slow gene that is expressed is not the same as the cardiac beta myosin gene (Jaenicke, Butterworth & Goldspink, 1992) as previously supposed (Lompre *et al.*, 1984). Indeed, it appears to be an embryonic or neonatal isoform slow gene and this is expressed first whenever skeletal muscle remodels. This has also been confirmed in other work we are doing with orthopaedic surgeons at Oxford, who are carrying out surgical limb lengthening; and with Dr Pamela Williams, who used a neonatal myosin hc antibody to stain sections of muscles that have been subjected to stretch far beyond the normal physiological limits (Fig. 4). Under these conditions the neonatal isoform is heavily expressed when the muscle is subjected to stretch (P. Williams, J. Kenwright, A. Simpson & G. Goldspink, unpublished data).

One of the main aims of our present work was to isolate and clone this intermediate myosin hc gene. This we have now achieved and we are in the process of characterising this gene (Jaenicke *et al.*, 1991). It seems not only to be one component in a series of switching of individual isoform genes during physiological adaptation, but it is also strongly expressed during the neonatal period (Jaenicke *et al.*, 1992), although only for a week or so before the adult genes are activated. The fact that it is a neonatal isoform gene indicates that the fibres may have to revert

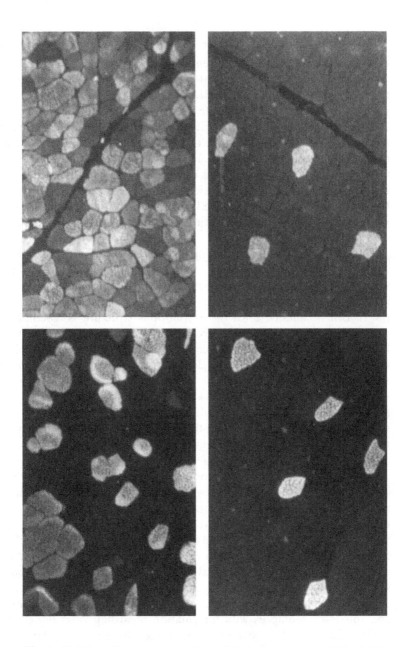

Fig. 4. Staining of transverse sections of (right) stretched, rabbit tibialis anterior during surgical limb lengthening with the same segment (left) of its contralateral control using rabbit monoclonal antibodies to (top) slow type I and (below) neonatal skeletal myosin heavy chains after one month. The antibody was donated by Dr Wendy Brown. From P. Williams, J. Kenwright & A. Simpson (unpublished data).

to expressing this gene before remodelling can occur. This is not a new concept as protein studies indicated that there was a required order for myosin gene switching (Butler-Brown & Whalen, 1984). The myosin hc proteins are difficult to separate electrophorectically because of their size, their low solubility and the fact that the isoforms have similar mass and charge. Therefore it is difficult to distinguish the neonatal and embryonic forms from the adult type IIa and IIb isoforms. However, we are more hopeful that we can do this using unique DNA sequences that are isoform-specific for *in situ* hybridisation to detect changes in mRNA levels. Also, as we are interested in the control of expression of these isoform genes, we need the flanking sequences where the regulatory elements are located.

The order and time required for the expression of individual genes of the myosin hc gene family may relate to their position on the chromosomes. Another suggested function for the neonatal/embryonic isoforms in skeletal muscle is that their more basic structure allows assembly of the myosin filaments; once the myofibrillar structure is assembled the neonatal/embryonic molecules are replaced by the predominant myosin hc type merely by law of mass action exchange. It is also possible that the embryonic and neonatal skeletal muscle myosin hc isoform genes exist because their regulation does not involve mechanical signals which would not be expected to have a significant role in early development and growth. In contrast, the adult isoform genes are required to endow mammalian skeletal muscle with the means of adapting to different contractile regimes imposed on the tissue.

We have the 5′ end, including the regulatory elements and unique coding sequences for this new gene and for the adult fast and slow genes. The promoter region of the new gene is particularly interesting as it represents an easily inducible promoter; it needs to be more fully characterised using cell transfection studies combined with and without cell stretching and compared with the adult isoform genes. By using different 5′ flanking fragments spliced to reporter genes, we can discern which elements impart developmental stage, cell type specific expression, and are responsive to changes in mechanical signals. As far as the coding region is concerned, *in situ* hybridisation is working well in the laboratory and this is providing us with information regarding what muscle fibres and at what stage this gene and other myosin hc genes are expressed.

The interconvertability of fibre types was demonstrated by cross-innervation and by chronic stimulation and has been reviewed (Pette & Vrbova, 1986; see also Fig. 5). It became generally accepted that the frequency of stimulation was the important factor in determining fibre type transition. However, more recently it was shown that higher stimula-

(*a*) Chromosome
in man

Adult skeletal genes

17.   emb.Myhc ➔ neo.MyHC ➔ MyHCIIb ➔ MyHCIIa

14.   ➔ MyHCb ↔ MyHCa

Cardiac genes

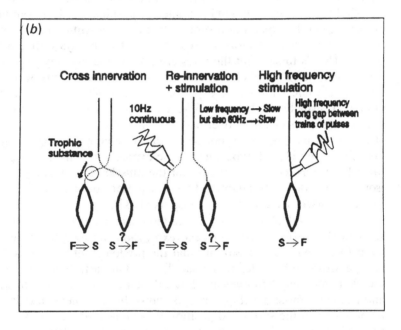

Fig. 5. (*a*) Arrangement of the skeletal and cardiac myosin hc genes on the different chromosomes in man. (*b*) Diagrammatic summary of previous studies of muscle fibre–phenotype transition induced by cross-innervation and electrical stimulation. Note that the transition from fast to slow can be induced by stimulation and that this does not have to be slow frequency stimulation as was previously proposed. Also, the conversion of slow to fast is more problematic and the type of stimulation that apparently results in a switch from the slow to the fast phenotype involves high frequency intermittent stimulation. It is proposed that the gaps in stimulation may be relevant as the trains of pulses may not be long enough to result in activation of the slow genes. Our work shows that both stretch and electrical stimulation both result in activation of the slow and repression of the fast genes. Lack of stretch and contractile activity result in the default expression of fast genes as shown in Fig. 6.

tion frequencies were just as effective in producing the fast to slow switch (Streter *et al.*, 1982). Certainly in our experiments, more complete reprogramming of the muscle was obtained when stretch was combined with high frequency stimulation. This indicates that the signal for the fast to slow change is mechanical strain rather than stimulation frequency *per se*.

Our work differs from the use of chronic stimulation in that protein synthesis is activated by stretch and by stimulation (Goldspink & Goldspink, 1986), because the muscle is rapidly producing new actin and myosin filaments in series and in parallel (Goldspink, 1984). In our model the change in the type of new protein that is being synthesised in abundant quantities can therefore be readily detected in days rather than in weeks. That is to say that the rates are not limited, as they presumably are in chronically stimulated muscle, to the protein turnover rates. Indeed, chronic stimulation experiments, in which the muscle is not stretched, usually result in atrophy. The experiments demonstrate that the isometric mechanical overload of a muscle, that is developed by stretch or force generation, not only induced rapid hypertrophy but also elicits slow muscle cell type–phenotypic expression. This makes physiological sense as it can be argued that the muscle cells by responding to isometric overload are adapting to a new pattern of activity that is akin to an increased postural role (Goldspink, 1985).

As mentioned above, protein studies have indicated that there is sequential myosin hc isoform gene expression during muscle development (Mahdavi *et al.*, 1986) and that the predominant protein isoforms change from embryonic, to neonatal, to adult fast. It seems that all muscle fibres stay phenotypically fast unless they are subjected to stretch and isometric force development. As shown by the soleus muscle when immobilised in the shortened position (Loughna *et al.*, 1986), subjected to surgical overload (Gregory *et al.*, 1986) or subjected to hypogravity (Oganov & Potapov, 1976), the muscle reverts to expressing the fast myosin genes unless it is receiving these mechanical stimuli (Fig. 6), under which circumstances the fast myosin genes are repressed. When the muscle is not subjected to stretch or force generation the fast myosin genes are expressed by default, which is essentially the same conclusion drawn by Swynghedauw (1986). He reviewed the conditions under which the fast myosin hc gene is expressed and refers to the fast type gene as the endogenous gene; but we prefer the term the 'default gene'. Certainly, the main regulation in the expression of the slow phenotype seems to depend on the repression of the fast type as much as the activation of the slow type genes. Other subsets of genes may be controlled in a similar fashion including mitochondrial and cytoplasmic enzyme genes. The changes in gene expression may not be coordinated and indeed there

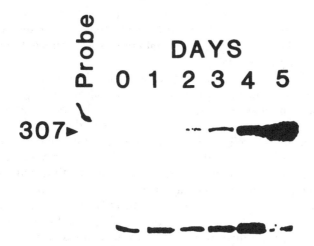

Fig. 6. Expression of the fast glycolytic (type IIb) myosin heavy chain gene in the rat soleus subjected to immobilisation in the shortened position for up to 5 days. The probe used was a single-stranded 304 nucleotide 3' end of pMHC 62. S1 nuclease digestion was carried out following hybridisation to remove any single-stranded RNA or DNA. This muscle, which does not normally express the fast glycolytic (type 116) gene, begins to express after 2 days when it is denied stretch and contractile activity. (From Loughna *et al.*, 1986, with permission.)

may be different signals involved, for example in the activation of the myosin isoform genes as compared to mitochondrial genes or sarcoplasmic enzyme genes, but under most training conditions these happen to coincide. For example, Williams *et al.* (1986) reported that 10 Hz chronic stimulation of the rabbit tibialis anterior muscle for 21 days resulted in a five-fold increase in cytochrome *b* mRNA but a four-fold reduction in the levels of aldolase mRNA. In the reverse direction, when muscle fibres adapt for increased power output rather than economy, there is also an increase in muscle mass. Hence the question is, why do sprinters develop large muscles which obviously remain fast contracting? This is probably explained by the finding that the very slow adult type I gene, which is the same as the cardiac beta myosin hc gene, needs applied stretch or repetitive contractile activity of long duration. The time window for inducing fibre hypertrophy seems somewhat shorter so that it is possible with the correct type of training, i.e. short intensive bursts, to induce hypertrophy but not the full fast to slow transition. Certainly, it is known from EMG studies that postural muscle fibres such as those of the soleus are activated about 90% of the time during standing or walking whilst the fibres in other skeletal muscles are activated only 5% of the time (Hnik *et al.*, 1985).

The details of the molecular mechanism(s) involved in isoform gene switching are not known. Two possible mechanisms spring to mind, including transient changes in internal calcium levels, and metabolic signals such as the depletion of ATP and the change in the phosphorylation potential of the muscle cell. If calcium is involved in the cellular signalling that activates muscle genes, the system may be frequency rather than amplitude dependent and there may be a minimum time required to activate some genes, but others may have a shorter time window. With respect to metabolic signals, Moerland, Wolfe & Kushmerick (1989) have shown that administration of a creatine analogue induces significant changes in the type of myosin protein isoforms expressed in the mouse soleus and extensor digitorum longus muscles. The magnitude and rapidity of the switch in isoform gene transcription and the synthesis of much new protein make the stretch/stimulation model a good system for investigating the mechanical signals, second messengers and transcription factors involved in muscle fibre–phenotype determination as well as in the control of muscle cell growth. The regulatory or early genes that code for putative transcription factor proteins such as c-*fos* and c-*myc* are also being studied (Izumo, Nadal-Ginard & Mahdavi, 1988), but we still need to know what regulates the regulators.

As an adaptive mechanism, the switch in expression from the fast to the slow type genes under conditions of static overload or repetitive contraction makes physiological sense. It can be looked on as an adaptation for increased economy as the slow myosin hc has a lower specific ATPase activity and uses less energy in maintaining isometric force and for slow repetitive movements (Goldspink, 1985). Further work is necessary, using more refined techniques in physiology and molecular biology to define the nature of the link between the mechanical signals and the activation and repression of subsets of muscle genes.

This work was funded by grants from NASA USA, the Wellcome Trust, and the AFRC.

References

Brownson, C., Isenberg, H., Brown, W. & Edwards, Y. (1989). Changes in skeletal muscle gene transcription induced by chronic stimulation. *Muscle and Nerve* **11**, 1183–9.

Butler-Brown, G. S. & Whalen, R. G. (1984). Myosin isozyme transitions occurring during the post-natal development of the rat soleus muscle. *Developmental Biology* **102**, 324–34.

Gerlach, G.-F., Turay, L., Malik, K., Lida, J., Scutt, A. & Goldspink, G. (1990). The mechanisms of seasonal temperature acclimation in the carp: a combined physiological and molecular biology approach. *American Journal of Physiology* **259**, R237–44.

Goldspink, G. (1984). Alterations in myofibril size and structure during growth, exercise and changes in environmental temperature. In *Handbook of Physiology – Skeletal Muscle*, ed. L. D. Peachy, R. H. Adrian & S. R. Geiger, pp. 593–654. American Physiological Society.

Goldspink, G. (1985). Malleability of the motor system: a comparative approach. In *Design and Performance of Muscular Systems*, ed. C. R. Taylor, E. Weibel & L. Bolis. *Journal of Exprimental Biology* **115**, 375–91.

Goldspink, D. F. & Goldspink, G. (1986). The role of passive stretch in retarding muscle atrophy. In *Electrical Stimulation and Neuromuscular Disorders*, ed. W. A. Nix & G. Vrbova, pp. 91–100. Berlin: Springer-Verlag.

Goldspink, G. & Griffin, G. (1973). The increase in skeletal muscle mass in male and female mice during post-natal growth. *Anatomical Record* **177**, 465–70.

Goldspink, G., Scutt, A., Loughna, P., Wells, D. Jaenicke, T. & Gerlach G.-F. (1992). Gene expression in skeletal muscle in response to mechanical signals. *American Journal of Physiology* **262**, R326–63.

Gregory, P., Low, R. & Stirewalt, W. S. (1986). Changes in skeletal muscle myosin isoenzymes with hypertrophy and exercise. *Biochemical Journal* **238**, 55–63.

Griffin, G., Williams, P. E. & Goldspink, G. (1971). Region of longitudinal growth in striated muscle fibres. *Nature, New Biology* **232** (27), 28–9.

Hnik, P., Vejsada, R., Goldspink, D. F., Kasiciki, S. & Krekule, I. (1985). Quantitative evaluation of EMG activity in rat extensor and flexor muscles immobilized at different lengths. *Experimental Neurology* **88**, 515–28.

Izumo, S., Nadal-Ginard, B. & Mahdavi, V. J. (1986). All members of the MHC multigene family respond to thyroid hormone in a highly tissue-specific manner. *Science* **231**, 597–600.

Izumo, S., Nadal-Ginard, B. & Mahdavi, V. (1988). Proto-oncogene induction and reprogramming of cardiac gene expression produced by pressure overload. *Proceedings of the National Academy of Science USA* **85** 339–43.

Jaenicke, T., Martindale, J., Scutt, A., Gerlach, G.-F., Loughna, P. F., Williams, P. & Goldspink, G. (1991). Mechanical signals involved in the expression of slow type genes and the determination of muscle-fibre phenotype in mammalian muscle. In *Muscle and Motility*, Vol. 2, ed. G. Marechal & U. Carraro. Andover, UK: Intercept.

Jaenicke, T., Butterworth, P. H. W. & Goldspink, G. (1992). Isolation and characterisation of a developmentally controlled myosin heavy chain. *Proceedings of the XXIth European Muscle Conference, Bielefeld*. Andover, UK: Intercept.

Lompre, A. M., Nadal-Ginard, B. & Mahdavi, V. J. (1984). Expression of the cardiac ventricular alpha and beta myosin heavy chain genes is developmentally and hormonally regulated. *Journal of Biological Chemistry* **259**, 6437–46.

Loughna, P., Goldspink, D. F. & Goldspink, G. (1986). Effect of inactivity and passive stretch on protein turnover in phasic and postural rat muscles. *Journal of Applied Physiology* **61**, 173–9.

Maeda, K., Sckziel, G. & Whittinghoffer, G. (1987). Characterisation of cDNA coding for the couple light meromyosin portion of a rabbit fast skeletal muscle myosin heavy chain. *European Journal of Biochemistry* **167**, 97–102.

Mahdavi, V., Strehler, E., Periasamy, M., Wieczyorek, D. E., Izumo, S. & Nadal-Ginard, B. (1986). Sarcomeric myosin heavy chain gene family: organisation and pattern of expression. *Medical Science of Sports and Exercise* **18**, 229–308.

Moerland, T. S., Wolfe, N. G. & Kushmerick, M. J. (1989). Administration of a creatine analogue induces isomyosin transitions in muscle. *American Journal of Physiology* **257** (Cell Physiol. 26), C810–16.

Oganov, V. S. & Potapov, A. N. (1976). On the mechanism of the changes in skeletal muscles in the weightless environment. *Life Science Space Research* **19**, 137–43.

Pette, D. & Vrbova, G. (1986). Neural control of phenotypic expression in mammalian muscle fibres. *Muscle and Nerve* **8**, 676–89.

Reiser, P. J., Moss, R. L., Giulian, G. C. & Greaser, M. L. (1985). Shortening velocity and myosin heavy chains of developing rabbit muscle fibres. *Journal of Biological Chemistry* **206**, 14403–5.

Streter, F. A., Pinter, K., Jolesz, F. & Mabauchi, K. (1982). Fast to slow transformation of fast muscles in response to long-term phasic stimulation. *Experimental Neurology* **75**, 95–102.

Swynghedauw, D. (1986). Developmental and functional adaptations of contractile proteins in cardiac and skeletal muscles. *Physiological Reviews* **65**, 710–71.

Umeda, P. K., Sinha, A. M., Jakovcic, S., Meren, S., Hsu, H.-J., Subramanian, K. N. & Zak, R. (1981). Molecular cloning of two fast myosin heavy chain cDNAs from chicken embryo skeletal muscle. *Proceedings of the National Academy of Sciences USA* **78**, 2843–7.

Weydert, A., Barton, P., Harris, A. J., Pinset, C. & Buckingham, M. (1987). Developmental pattern of mouse skeletal myosin heavy chain gene transcripts *in vivo* and *in vitro*. *Cell* **49**, 121–9.

Williams, P. E. & Goldspink, G. (1973). The effect of immobilization on the longitudinal growth of stretch and muscle fibres. *Journal of Anatomy* **116**, 45–55.

Williams, P. E., Watt, P., Bicik, P. & Goldspink, G. (1986). Effect of stretch combined with electrical stimulation on the type of sarcomeres produced at the ends of muscle fibres. *Experimental Neurology* **93**, 500–9.

Williams, R. S., Salmons, R., Newsholme, A., Kaufman, R. E. & Mellor, J. (1986). Regulation of nuclear and mitochondrial gene expression by contractile activity in skeletal muscle. *Journal of Biological Chemistry* **261**, 376–80.

S. E. BLACKSHAW

Stretch sensitivity in stretch receptor neurones

Introduction

The sensory terminals of stretch receptor neurones are designed to transduce the energy of a mechanical stimulus into an electrical signal. The
molecular mechanisms by which these primary sensory cells sense physical forces are not known. It is generally assumed that depolarisation of
the sensory terminals is the result of the activation of ion channels that
are gated by mechanical stimuli. But direct characterisation of mechanosensitive channels in these specialised sensory cells using patch clamp
techniques has been limited because of the inaccessibility of the sensory
terminals. Stretch-activated channels have been studied in a number of
other cell types that are not specialised mechanosensory cells, but
whether such channels are good models for mechanosensitive channels
of mechanoreceptors is not clear. Stretch receptor neurones in the medi-
cinal leech are particularly useful for studies of mechanosensory transduction because they have large unbranched sensory terminals adjacent
to accessible cell bodies.

Electrophysiological studies of vertebrate spindles and crustacean stretch receptors

In the 1950s Katz showed, by recording with extracellular electrodes
close to the sensory ending in frog muscle spindles during stretch, that
the first detectable step in the transduction process was a local depolarisation of the sensory terminals, termed the receptor potential. Katz used
isolated frog muscle spindles that contain between three and 12 intrafusal
muscle fibres innervated by a single sensory axon that branches
repeatedly and coils around the intrafusal fibres (Katz, 1950). The sensory cell bodies in these receptors are some distance away in the dorsal
root ganglia adjacent to the spinal cord. Recordings were made with

Society for Experimental Biology Seminar Series 54: *Biomechanics and Cells*, ed. F. Lyall &
A. J. El Haj. © Cambridge University Press 1994, pp. 96–106.

external electrodes close to the sensory ending but not in direct contact. Katz showed that there were both dynamic and static components to the receptor potential: an initial peak which coincided with the period of dynamic lengthening of the muscle, and a later static component during maintained stretch in which the local depolarisation was maintained at a lower level (Katz, 1950). Katz's conclusion was that the receptor potential was closely related in its time course and amplitude to the mechanical stimulus, and that it was presumably the direct result of the deformation of the nerve endings and their receptor membranes. The fact that the receptor potential could be recorded in preparations in which spiking activity in the sensory terminal was blocked with local anaesthetic suggested that the discharge of action potentials was a secondary phenomenon, initiated by the receptor potential. In these experiments, and in subsequent experiments on isolated mammalian spindles (reviewed in Hunt, 1990; Blackshaw, 1992) the receptor potential was recorded externally. In Katz's experiments, the recording electrode was outside the spindle capsule, i.e. close to the sensory terminal but not in direct contact, and because of electronic spread along the fine terminal branches of the sensory ending, the signal recorded was necessarily a reduced image of events at the site of transduction. Neither intracellular recordings nor single channel recordings have been possible from sensory terminals in spindles because of the inaccessibility and small diameter of the terminal branches.

Stretch receptor neurones in a different preparation proved to be particularly well suited to electrophysiological study of mechanotransduction: the primary sensory neurones in the abdominal muscles of crayfish, recognised by Alexandrowicz (1951, 1967). In this preparation the sensory cell body lies adjacent to the receptor muscle in the periphery and projects its dendrites into the muscle. Eyzaguirre & Kuffler (1955) used a preparation consisting of a receptor muscle separated from the dorsal musculature with its sensory innervation intact. The natural insertions of the receptor muscle into the exoskeleton were maintained and stretch was applied by pulling simultaneously on the shell at both ends of the receptor muscle. The great technical advantage of this preparation is that when the living isolated structure is viewed with dark field microscopy it is possible to visualise cellular details such as the nucleus and parts of the dendrites, and as a consequence to position electrodes accurately and to record intracellularly close to the transducing membrane of the dendrites. By using flexible microelectrodes they were able to maintain an intracellular electrode in the sensory cell body during stretch of the receptor muscle and to make the first intracellular recordings of membrane potential changes during stretch.

Their recordings (Eyzaguirre & Kuffler, 1955) of receptor potentials in the presence of novocaine to block spiking activity show a decline in the depolarisation during the initial period of maintained stretch to a lower maintained value which corresponded to the dynamic and static phases of the sensory responses recorded by Katz from frog muscle spindles. In these experiments recordings were made from the cell body or large branches of the dendrites. It has not been possible to record intracellularly from the fine dendritic terminals which are presumed to be where the receptor potential is generated. As in the spindle experiments, the assumption is that the subthreshold graded potentials measured reflect the changing levels of the receptor potential at its site of origin in the dendrites. The actual amount of stretch depolarisation in the dendritic terminals is not known but it is possible to make estimates by using values for space constant obtained in other crustacean nerve fibres, assuming spatial decrement of the signal, and allowing for the smaller diameter of the dendrites.

Subsequent voltage clamp and ion substitution experiments to determine the ionic basis of the receptor potential showed that the sensory membrane became permeable to sodium and other cation species during stretch, thus accounting for the depolarisation (Terzuolo & Washizu, 1962; Edwards, Terzuolo & Washizu, 1963; Obara, 1968; Obara & Grundfest, 1968; Klie & Wellhoner, 1973; Brown et al., 1978; Edwards et al., 1981). Similarly ion substitution experiments in vertebrate spindles showed that the receptor potential is associated with a non-selective increase in cation permeability (Ottoson, 1964; Albuquerque, Chung & Ottoson, 1969; Hunt, Wilkinson & Fukami, 1978).

Mechanosensitive ion channels

The detailed mechanism by which mechanical distortion produces a change in permeability in these cells is not known. One of the consequences of the development of single channel recording techniques has been the discovery of ion channels that are activated by stretch. Mechanosensitive ion channels were discovered in 1984, some 30 years after Katz's first recording of the receptor potential in frog spindle afferents (Brehm, Kullberg & Moody-Corbett, 1984; Guharay & Sachs, 1984). They were found not in mechanosensory neurones, but in muscle cell membranes, and they have since been reported in a variety of cell types including smooth muscle, skeletal muscle, nerve and bacteria, indicating that mechanosensitivity is a common property of cell membranes (for reviews see Sachs, 1988, 1989; Morris, 1990, 1992). Mechanosensitive channels in cells other than specialised mechanosensory cells have been implicated in a number of basic physiological functions, including the regulation of cell volume. Channels sensitive to cell swelling are found not only in cells such as kidney tubule cells that have to cope with osmotic stress and where volume regulation is considered to be part of their normal functioning, but also in cells which are not supposed to face an osmotic challenge under physiological conditions, for example neuroblastoma cells or bovine chromaffin cells, possibly indicating that volume-sensitive ion channels are distributed more widely than previously appreciated (Morris, 1990, 1992). Cation selective channels with a conductance of 40 pS have been demonstrated in vascular endothelial cells cultured from aorta (Lansman, Hallam & Rink, 1987). Mechanosensitive channels have also been recorded in brain capillary endothelial cells (Popp *et al.*, 1992) and may act as mechanotransducers for haemodynamic stress in this kind of cell.

Doroshenko & Neher (1992) suggest that two major types of volume-sensitive ion channels exist, differing in their relationship to the physical stress applied to the cell membrane: channels whose activation is directly linked to the magnitude of the applied stress (Morris, 1990), and also channels for which induced deformations of the cell membrane and its cytoskeleton can play only the role of a trigger (e.g. VS chloride channels in bovine chromaffin cells). Both types of volume-sensitive channels may coexist within the same cell type.

Stretch activated ion channels in sensory neurones

Direct characterisation of mechanosensitive ion channels in specialised mechanosensory neurones has been limited for technical reasons, because of the difficulty of access to the transducing regions of the cells. It is in the nature of mechanosensory neurones that the terminal dendrites are intimately associated with muscle or connective tissue and consequently inaccessible to patch electrodes. There has to date been only one single channel study of mechanosensitive ion channels in stretch receptor neurones, made by Erxleben on the crustacean abdominal receptor (Erxleben, 1989). In these experiments recordings were made from the soma and primary dendrites and not from the terminal dendritic branches. Two types of channel activated by membrane stretch were seen, both permeable to cations: these were indistinguishable on the basis of their ion selectivity of single channel conductance, but differed in their sensitivity to membrane deformation and voltage range of activation. The first type of channel, seen mainly in cell-body or axon hillock patches, was largely inactive if no suction was applied to the patch. The probability of this channel being open increased with hyperpolarisation and this channel is referred to as the inwardly rectifying stretch-activated (RSA) channel. It was not clear what contribution this channel might make to the generator current: the density of channels was low and they were only active at rather negative potentials. The second type of channel, referred to as the stretch-activated (SA) channel had properties appropriate for the mechanotransduction channel. It was seen in large numbers in patches from the primary dendrites; the open probability of the channel increased with increasing stretch applied to the patch, and it had the low cation selectivity established for the receptor current in previous voltage clamp and ion substitution experiments. Current voltage relationships were approximately linear with slope conductances for K^+ of 71 ± 11 pS (SD, $n = 3$); for Na^+ of 50 ± 7.4 pS ($n = 5$) and for Ca^{2+} of 22 ± 3 pS ($n = 3$). This channel was active in the absence of suction (mean Po RSA: 9×10^{-4}, mean Po SA: 1.3×10^{-2}) and was largely voltage-independent.

What other evidence is there for SA ion channels in mechanosensory neurones? Single-channel recordings of stretch-activated ion channels have been made from cutaneous mechanosensory neurones in the leech and *Aplysia*. Recordings were made in *Aplysia* from cultured mechanosensory neurones of the ventro-medial cluster of the pleural ganglion that respond to peripheral mechanostimulation (Vandorpe & Morris, 1991); in the leech from cutaneous mechanosensory neurones that respond to light touch or noxious stimuli to the skin of the leech (Blackshaw, 1981; Blackshaw, Nicholls & Parnas, 1982; Pellegrino *et al.*, 1990). In both these studies recordings were made from the neurone cell bodies which are situated centrally within the ganglion at some distance from the sensory terminals where mechanotransduction takes place, and it is unlikely that these cell body channels play a direct role in mechanoexcitation. In both leech and *Aplysia* the stretch-activated channels in cell body membrane were potassium-selective, unlike the non-selective cation permeability underlying the receptor potential of stretch receptor neurones. Vandorpe & Morris (1991) thought the *Aplysia* channels likely to be related to stretch-activated K channels widely found in snail neurones in culture (Sigurdson & Morris, 1989).

Stretch-activated K channels have also been recorded from the lateral wall of auditory outer hair cells in the vertebrate acoustico-lateralis system (Iwasa *et al.*, 1991). These channels are K-selective and therefore different from the non-selective channels activated by deflection of the stereocilia of hair cells (Hudspeth, 1989). The lateral wall of the outer hair cell is estimated to contain about 95 channels, i.e. they are present at a relatively low density but comparable to the estimated total number of mechanotransducer channels associated with the stereocilia. Iwasa *et al.* (1991) suggest that the elongated cell bodies of the outer hair cells are optimally positioned to sense axial forces due to vibrations of the basilar membrane during sound stimulation, and that acoustic stimuli stretch the outer hair cell as well as deflecting the stereocilia. Outer hair cells are thought to have a motor function (Ashmore, 1993). They change length when stimulated electrically and these length changes can occur at acoustic frequencies. Their ability to generate forces at high frequency means that they are well placed to control local mechanical properties of the basilar membrane. The stretch-sensitive channels in the lateral wall of the outer hair cells may play a role in detecting length changes in the hair cells.

Stretch receptor neurones in leech

Stretch receptor neurones in leech are particularly useful for studies of mechanosensory transduction because, as in the crustacean abdominal receptors, the cell bodies lie singly in the periphery. In the leech the cell bodies occur at predictable locations along specific nerve roots and are adjacent to the sensory dendrites. They can be visualised in live preparations and they are large enough for penetration by intracellular electrodes. These neurones provide the sensory innervation for the tubular muscle layers of the leech body wall (reviewed in Blackshaw, 1993).

The fact that the cell bodies are large and accessible makes it possible to label the cells via microelectrodes inserted into the cell body with tracers such as horseradish peroxidase, or with fluorescently labelled antibodies. The sensory dendrites are large, flat, fan-shaped structures about 70 μm across but only a few micrometers deep. Unlike the sensory terminals of spindle afferents or crustacean abdominal receptors, the leech sensory dendrites do not branch but are single large structures associated with muscle at their lateral edges. Each neurone has two of these flattened dendrites, one is at the end of the peripheral axon, adjacent to the cell body, and associated with a band of longitudinal muscle fibres; the second dendrite is a flattened 'en passant' expansion of the axon, located between the cell body and the ganglion, and associated with a different longitudinal muscle band from the distal dendrite. This gives an arrangement whereby two transducing regions of the cell are arranged in series, a few hundred micrometres apart and separated by the cell body.

The sensory axons project into the segmental ganglia of the ventral nerve cord. At 10 μm diameter they are the largest axons in the leech nervous system. They can be seen within the nerve roots using dark field illumination and intracellular recordings can be made from different regions of the axon, either adjacent to the cell body and dendrites or close to the CNS (Blackshaw & Thompson, 1988). Direct injection of current into the cell body or into different regions of the axon shows that action potentials can be elicited from the peripheral cell body of the neurone but not from the axon, i.e. imposed voltage changes spread decrementally along the axon. The space constant of these fibres, estimated from the decrement of a steady state potential between two points along the axon, is between 3 and 5 mm, consistent with the requirements of a decrementally conducting neurone.

The sensory responses to stretch of these neurones are unusual. They have been studied by recording intracellularly from the axon while stretching restricted areas of longitudinal muscle associated with the dendrites. Ramp stretch of the muscle elicits hyperpolarising potentials that are maintained for the duration of the stretch, whose amplitude depends on the extent of the final displacement. Release from stretch elicits depolarising responses, often characterised by small spike-like events. The excitatory responses to release consist of an initial dynamic component followed by a smaller amplitude static component maintained for the duration of the ramp stimulus. The amplitude of both dynamic and static components depends on the extent of the final displacement, while the rate of rise of the dynamic component depends on the velocity of release. Simultaneous measurement of tension within the muscle during ramp release shows that the leech receptors may more closely reflect tension changes within the muscle rather than length *per se*.

In summary, the leech muscle receptors have expanded dendrites associated with the longitudinal muscle of the body wall, and large axons projecting into the CNS. They transduce a stretch stimulus to muscle, but the sensory responses to stretch are unusual in being bi-directional, i.e. these neurones are hyperpolarised by muscle stretch and depolarised by release from stretch. In this respect they differ from the better known vertebrate spindles or crustacean stretch receptors which are depolarised by their adequate stimulus, and more closely resemble mechanoreceptive hair cells in the vertebrate acoustico-lateralis system which have both hyperpolarising and depolarising receptor potentials depending on which way the hair is deflected. We do not yet know whether the hyperpolarising responses to stretch are the result of an increase or decrease in conductance of the transducing membrane, nor what the ionic basis of the receptor potential is in this unusual receptor. It is noteworthy that all mechanosensitive channels of mechanosensory neurones so far studied are cationic channels, whereas some mechanosensitive channels in non-specialised sensory cells are permeable to anions. For example, rabbit cardiac myocytes show stretch-activated anion currents (Hagiwara *et al.*,

1992) and there is an increase in chloride conductance on cell inflation in bovine chromaffin cells (Doroshenko & Neher, 1992).

Unusual ultrastructural features of leech sensory terminals

Light and electron microscope studies of the dendrites of the leech receptors show that their morphology as well as their sensory responses is unusual. When the dendrites are viewed with DIC optics in whole mounts of the body wall, one surface of the dendrite has a distinctive knobbly appearance. Fine sections show that the two opposing surfaces of the expanded dendrite have very different appearances, one highly invaginated, one a relatively flat membrane. The knobbly appearance in the light microscope is due to numerous small protuberances (approximately 1 μm in diameter) attached via slender stalks to the flat surface of the dendrite. These protuberances have an internal structure consisting of concentric layers of membrane-bound cytoplasm separated by extracellular space (Blackshaw, McKay & Thompson, 1984). We do not know what the function of these small protuberances is. They may be a mechanism for increasing the surface area of dendritic membrane – perhaps in order to accommodate large numbers of mechanosensitive ion channels at appropriate sites. Or perhaps in a highly extensible animal like the leech there needs to be a source of spare membrane to accommodate stretch.

Concluding remarks

There has been limited analysis of mechanosensitive channels in mechanosensory neurones and the molecular mechanisms by which these cells sense physical forces are not well understood. More single-channel studies of stretch receptor neurones are needed. The dissected mammalian spindle or crustacean receptor may not be best suited as preparations for studying the molecular mechanisms by which mechanical distortion produces a change in permeability in sensory terminals. The molecular characterisation of stretch receptor channels might be best approached in preparations like the leech which have large unbranched sensory terminals, or where chosen cells can be removed for single-channel study in culture or molecular genetic studies. The cell biological aspects of force transmission are also not well understood. The type of deformation produced by stretch, the molecular basis of attachment sites between sensory and muscle membranes, and the role of the extracellular matrix are all largely uncharacterised. Little is also known about the cyto-

104   S. E. BLACKSHAW

skeleton in stretch receptor neurones or about the response of the cytoskeleton to imposed shape changes, or whether links between the extracellular matrix and the cell's cytoskeleton are important for force
transduction.

Acknowledgements

Experiments on leech stretch receptors were supported by the SERC and
MRC.

References

Albuquerque, E. X., Chung, S. H. & Ottoson, D. (1969). The action
of tetrodotoxin on the frog's isolated spindle. *Acta Physiologica Scan-
dinavica* **75**, 301–12.
Alexandrowicz, J. S. (1951). Muscle receptor organ in the abdomen of
Homarus vulgaris and *Palinurus vulgaris*. *Quarterly Journal of Microscopic Science* **92**, 163–99.
Alexandrowicz, J. S. (1967). Receptor organs in the thoracic and
abdominal muscle of Crustacea. *Biological Reviews* **42**, 288–326.
Ashmore, J. (1993). The ear's fast cellular motor. *Current Biology* **3**
(1), 38–40.
Blackshaw, S. E. (1981). Morphology and distribution of touch cell
terminals in the skin of the leech. *Journal of Physiology* **320**, 219–28.
Blackshaw, S. E. (1992). Experimental approaches to transduction and
the receptor potential in muscle receptors. *Progress in Biophysics and
Molecular Biology* **58**, 19–60.
Blackshaw, S. E. (1993). Stretch receptors and body wall muscle in
leeches. *Comparative Biochemistry and Physiology* **105A**, 643–52.
Blackshaw, S. E., McKay, D. A. & Thompson, S. W. N. (1984). The
fine structure of leech stretch receptor neurone and its efferent supply.
Journal of Physiology **350**, 76P.
Blackshaw, S. E., Nicholls, J. G. & Parnas, I. (1982). Physiological
responses, receptive fields and terminal arborisations of nociceptive
cells in the leech. *Journal of Physiology* **326**, 251–60.
Blackshaw, S. E. & Thompson, S. W. N. (1988). Hyperpolarising
responses to stretch in neurones innervating leech body wall muscle.
Journal of Physiology **396**, 121–38.
Brehm, P., Kullberg, R. & Moody-Corbett, F. (1984). Properties of
non-junctional acetylcholine receptor channels on innervated muscle
of *Xenopus Laevis*. *Journal of Physiology* **350**, 631–48.
Brown, H. Mack, Ottoson, D. & Rydqvist, B. (1978). Crayfish stretch
receptor: an investigation with voltage-clamp and ion-sensitive electrodes. *Journal of Physiology* **284**, 155–79.

Doroshenko, P. & Neher, E. (1992). Voltage-sensitive chloride conductance in bovine chromaffin cell membrane. *Journal of Physiology* **449**, 197–219.

Edwards, C., Terzuolo, C. A. & Washizu, Y. (1963). The effect of changes of the ionic environment upon an isolated crustacean sensory neuron. *Journal of Neurophysiology* **26**, 948–57.

Edwards, C., Ottoson, D., Rydqvist, B. & Swerup, C. (1981). The permeability of the transducer membrane of the crayfish stretch receptor to calcium and other divalent cations. *Neuroscience* **6**, 1455–60.

Erxleben, C. (1989). Stretch-activated current through single ion channels in the abdominal stretch receptor organ of the crayfish. *Journal of general Physiology* **94**, 1071–83.

Eyzaguirre, C. & Kuffler, S. W. (1955). Processes of excitation in the dendrites and in the soma of single isolated sensory nerve cells of the lobster and crayfish. *Journal of Physiology* **39**, 87–119.

Guharay, F. & Sachs, F. (1984). Stretch activated single ion channel currents in tissue-cultured embryonic chick skeletal muscle. *Journal of Physiology* **352**, 685–701.

Hagiwara, N., Masuda, H., Shoda, M. & Irisawa, H. (1992). Stretch-activated anion currents of rabbit cardiac myocytes. *Journal of Physiology* **456**, 285–302.

Hudspeth, A. J. (1989). How the ear's works work. *Nature* **343**, 814–16.

Hunt, C. C. (1990). Mammalian muscle spindle: peripheral mechanisms. *Physiological Reviews* **70** (3), 643–63.

Hunt, C. C., Wilkinson, R. S. & Fukami, Y. (1978). Ionic basis of the receptor potential in primary endings of mammalian muscle spindles. *Journal of general Physiology* **71**, 683–98.

Iwasa, K. H., Li, M., Jia, M. & Kachar, B. (1991). Stretch sensitivity of the lateral wall of the auditory outer hair cell from the guinea pig. *Neuroscience Letters* **133**, 171–4.

Katz, B. (1950). Depolarisation of sensory terminals and the initiation of impulses in the muscle spindle. *Journal of Physiology* **111**, 261–82.

Klie, J. W. & Wellhoner, H. H. (1973). Voltage clamp studies on the stretch response in the neuron of the slowly adapting crayfish stretch receptor. *Pflügers Archiv* **342**, 93–104.

Lansman, J. B., Hallam, T. J. & Rink, T. J. (1987). Single stretch-activated ion channels in vascular endothelial cells as mechanotransducers? *Nature* **325**, 811–13.

Morris, C. E. (1990). Mechanosensitive ion channels. *Journal of Membrane Biology* **113**, 93–107.

Morris, C. E. (1992). Are stretch-sensitive channels in molluscan cells and elsewhere physiological transducers? *Experientia* **48**, 852–8.

Morris, C. E. & Sigurdson, W. J. (1989). Stretch-inactivated ion channels co-exist with stretch-activated ion channels. *Science* **243**, 807–9.

Obara, S. (1968). Effects of some organic cations on generator potential of crayfish stretch receptor. *Journal of general Physiology* **52**, 363–86.

Obara, S. & Grundfest, H. (1968). Effects of lithium on different membrane components in crayfish stretch receptor. *Journal of general Physiology* **51**, 635–54.

Ottoson, D. (1964). The effect of sodium deficiency on the response of the isolated muscle spindle. *Journal of Physiology* **171**, 109–18.

Pellegrino, M., Pellegrini, M., Simoni, A. & Gargini, C. (1990). Stretch-activated cation channels with large unitary conductance in leech central neurons. *Brain Research* **525**, 322–6.

Popp, R., Hoyer, J., Meyer, J., Galla, H.-J. & Gogelein, H. (1992). Stretch-activated non-selective cation channels in the antiluminal membrane of porcine cerebral capillaries. *Journal of Physiology* **454**, 435–49.

Sachs, F. (1988). Mechanical transduction in Biological Systems. *CRC Critical Reviews of Biomedical Engineering* **16**, 141–69.

Sachs, F. (1989). Ion channels as mechanical transducers. In *Cell Shape: Determinants, Regulation and Regulatory Role*, ed. W. D. Stein & F. Bronner, pp. 63–92. San Diego: Academic Press.

Sigurdson, W. J. & Morris, C. E. (1989). Stretch-activated ion channels in growth cores of snail neurones. *Journal of Neuroscience* **9**, 2801–8.

Sigurdson, W. J., Morris, C. E., Brezden, B. B. & Gardner, D. R. (1987). Stretch activation of a K^+ channel in molluscan heart cells. *Journal of experimental Biology* **127**, 191–209.

Terzuolo, C. A. & Washizu, Y. (1962). Relation between stimulus strength, generator potential and impulse frequency in stretch receptor of crustacea. *Journal of Neurophysiology* **25**, 56–66.

Vandorpe, D. H. & Morris, C. E. (1991). A stretch-sensitive K channel from cultured *Aplysia* mechanosensory neurons. *Biophysics Journal* **59**, 455a.

J. S. HESLOP-HARRISON

Mechanical interactions with plant cells: a selective overview

Introduction

The study of mechanical interactions between plant cells, and the effects of mechanical stimuli on plant cells, represent an active but fragmented research field which covers many disciplines, ranging from interactions at the level of the ecosystem – in plant communities and in pathogen interactions with crops – to thigmotropic responses and the molecular genetics of gene induction.

The plant cell is surrounded by a rigid wall which normally consists of cellulose and other polysaccharides, and may be thickened with lignin (wood, a complex cross-linked phenolic substance) or subarin (cork). The plasma membrane is contained within the wall. The wall itself generally prohibits the sort of morphogenetic movements that are commonplace in animal development and many of those that do occur involve interactions of two distinct cellular genomes with different developmental potentials (Sussex et al., 1985). Hence the range of interactions and perhaps their developmental importance are lower than in animals. Nevertheless, there are advantages in the study of plant systems which can offer possibilities of genetic manipulation and accessibility. As in all studies, interactions between cells involve not only mechanical stimuli, but electrical, hormonal, chemical and nutritional (including water and light) factors, and research is often hindered by the difficulty of separating the different stimuli.

Many aspects of intercellular interaction phenomena in plants are reviewed by Linskens & Heslop-Harrison (1984). A general source of information on the topic of plant cell–cell interactions is given in Sussex et al. (1985), although the individual summaries presented in that book cannot be cited as publications. The mechanical interactions between

Society for Experimental Biology Seminar Series 54: *Biomechanics and Cells*, ed. F. Lyall & A. J. El Haj. © Cambridge University Press 1994, pp. 107–120.

plants and fungi are discussed by several authors writing in Cole & Hoch (1991).

It is not my intention to review here all work in the area but to discuss selectively some recent work in two broad divisions of research: interactions between cells and external stimuli, and between two or more different cells. Cell–cell interactions include some of the most extensively studied responses in plants because of their economic importance. The interactions between the pollen grain or fungal spore involve interactions between single cell from one organism interacting with many cells from another. Both pollen tubes and fungal hyphae grow by tip growth, rather than extension of the wall or stretching as in root elongation or other growth processes. Some of the responses by the potential host plants are related to – or the same as – those of wound responses, and hence this major type of mechanical response will be discussed first.

Mechanical cell interactions: the wound response

Effects of wounding

One of the most extreme mechanical effects on cells is wounding: penetration of the wall and plasmalemma by a foreign object. Wounds are relatively straightforward to study in the laboratory, since the induction is defined and the response extreme. The plant cuticle, like the epithelial surfaces of vertebrates, represents the principal passive (mechanical) barrier to fungal invasion, and protects the plant from mechanical damage and insect attack. Nevertheless, many plant cells are challenged by wounding: from biting by insects or other animals, by a single fungal cell penetrating, or by mechanical damage from wind or other agents. Quite gentle stimuli, such as leaves brushing against each other, may inhibit leaf growth, and induce wound responses as surface hairs or trichomes are damaged. Such gentle mechanical interactions may be significant in limiting plant growth in dense communities.

Since the 1920s, it has been known that active masses of cytoplasm aggregate in individual cells at mechanically induced wound sites in leaf epidermal cells (Nichols, 1925). The materials may seal wounds within a few minutes so they can withstand turgor (Russo & Bushnell, 1989) and prevent loss of cytoplasm or ions, while synthesis of new compounds including callose and phenylpropanoids is induced rapidly (Aist, 1976). The materials that aggregate at the wound site can encase needles or electrodes placed in a cell (Nims, Halliwell & Rosberg, 1967).

Plants also respond to mechanical stress and pathogen attack by synthesising a variety of proteins that are thought to provide protection against further trauma, and to inhibit both the growth and spread of invasive pathogens (Stanford, Northcote & Bevan, 1990). For example, mRNAs for hydroxyproline-rich glycoproteins and for enzymes catalysing steps in phenylpropanoid metabolism (Stanford, Bevan & Northcote, 1989) accumulate following a rapid increase in transcription of the genes after stimulation. Genes involved in plant defence responses exhibit differences in magnitude, spatial pattern and timing of induction in response to stress. Some genes show highly localised induction, while others show systemic induction at a distance from the wound or simulation site.

Wound-induced genes

One of the best characterised wound-induced genes has been isolated from wounded potato tubers (Shirras & Northcote, 1984). The expression of the gene increases by orders of magnitude following wounding, and the control and activity of the promoter, one of the wound-induced *win* family of genes, has been studied in detail by Stanford *et al.* (1990). They fused the 5' flanking sequence of the gene *win2* to the widely used beta-glucuronidase (GUS) reporter gene, and a selectable marker, NPTII, which gives resistance to kanamycin and hence enables selection of transformed cells. The construct was then transformed into tobacco and potato cells, two closely related species in the Solanaceae, and regenerated into plants. As controls, a constitutively expressed promoter, 35SCaMV and a truncated *win2* flanking sequence, fused to GUS, were also transformed into plants. The GUS gene allows sites of expression of the enzyme to be detected by precipitation of a blue reaction product when the tissue is treated with a chromogenic substrate (Jefferson, Kavanagh & Bevan, 1987), and hence the construct enabled study of the timing of induction, the spatial patterns and the cell specificity of the transcription of genes in response to mechanical and biological stress.

Stanford *et al.* (1990) observed dramatic inductions of GUS expression upon wounding potato leaves, stems and tubers, with striking patterns of local and distant induction being found from 16 h after wounding in different organ types and at different developmental stages. In contrast, chimeric gene activity was not regulated in a wound-specific manner in leaves of transgenic tobacco. In potato, gene fusions containing the shortened *win2* promoter did not show expression before or after wounding (no inducibility), while the entire leaf showed activity with 35SCaMV the constitutive promoter.

This transgenic experiment has allowed detailed analysis of a wound-induced promoter, and similar techniques may be valuable for analysis of genes induced by more subtle mechanical stimuli. Although the *win* promoter is clearly activated by the wounding, it cannot be the sequence which directly responds to wounding or generalised effects of it, since the heterologous fusions of the gene into tobacco showed no wound inducible activity. Furthermore, the signal transduction mechanisms have not been studied; while loss of turgor pressure or solutes from the cell may be signals, it has been suggested (Darvill *et al.*, 1989) that fragments of the cell wall, oligosaccharins, might be released into the plant and induce expression of genes regulated by promoters such as *win2*.

Induction of gene expression and nuclear architecture

Mechanisms of gene induction clearly involve a reorganisation of gene expression within the nucleus. New genes are turned on, and the transcription factors must move. In the experiments discussed (Stanford *et al.*, 1990), 16 different, replicate transformed plants resistant to kanamycin were made using the *win2*–GUS construct in potato. Twelve of these showed wound inducibility of GUS expression, and the others were not analysed further; they may have had rearrangements of the construct or been inactive for other reasons. Of the remaining plants, with wound-inducible genes, five were analysed in detail, and differences in the amount and tissue specificity of GUS activity following wounding were found, varying by some five-fold in stems attached to wounded leaves. More dramatically, inducibility in different tissues varied between transformants. One plant showed negligible induction in the leaves, while induction in the stems was high. Two constructs showed virtually no inducibility in roots, while the others showed between three and 20-fold increases in gene expression. Although the differences might be explained by variation within replicate samples, differences in the degree or sites of mechanical wounding and developmental differences between leaves (Stanford *et al.*, 1990), it might also be an effect of the different genomic sites into which the construct has been integrated.

It is common for transgenes to show differential regulation, and in some cases, re-transformation of a plant may turn off expression of the first gene (Matzke & Matzke, 1990, 1991). Understanding the physical architecture of the genome within the nucleus has the potential to show how genes may move during expression, and how differences in genomic context of genes may affect their expression. In the human, modulation of centromere position both during the cell cycle (Bartholdi, 1991) and in disease syndromes (Borden & Manuelidis, 1988) has been demonstrated, and significant work by Lawrence and colleagues is now showing where genes are located during expression and where RNA transcript processing occurs (Lawrence & Singer, 1985; Lawrence, Singer & Marselle, 1989; Carter *et al.*, 1993; Xing *et al.*, 1993). Understanding high-level genome structure within individual interphase nuclei is important for analysis of gene behaviour and has the potential to show how gene expression might be modulated in response to changing demands. Investigations of mechanical effects on gene expression provide a valuable model for understanding the switching or activation of gene expression.

Mechanical interactions between plants and pathogens

Pathogen interactions

Understanding the interactions between the pathogens and the plants they infect is of great economic importance. At the single cell level, the interaction determines whether infection will occur, and hence whether disease will develop. Hence plant–pathogen behaviour has been the subject of considerable research, and the mechanical aspects of the interaction have not been overlooked (Cole & Hoch, 1991).

Many wound responses discussed above (p. 108) form part of a plant's response to infection: pathogens normally penetrate the host, and induce the wound responses. Two major fungal infections, which account for much of the work on mechanical interactions, are the powdery mildews, such as *Erysiphe graminis* on cereals, and the rusts, including *Uromyces appendiculatus* and *Puccinia graminis*. In both cases, the leaf surface is the first point of interaction between the two organisms.

The attachment of the fungal propagules to a host plant is essential to the successful establishment of the pathogen. Attachment may involve recognition of the host surface, and an active process of secretion of adhesive materials by the fungus, which in some cases may be highly specific to a particular host species (Nicholson & Epstein, 1991). As well as mechanical penetration, many fungi also actively breach the passive barriers of the cuticle with hydrolytic enzymes produced by the fungi which are associated with cuticular/epithelial penetration (Cole & Hoch, 1991). However, there have been relatively few studies comparing directly the structures produced at mechanical wound sites with structures at sites of attempted infection, which might separate the mechanical and chemical aspects of host penetration. Furthermore, many wounding experiments have been under non-sterile conditions, so secondary infections may have given spurious results (Dr P. Smith, personal communication). Russo & Bushnell (1989) compared wounds left by needles and from fungal infection of barley epidermal cells. A wound plug was formed as a wall apposition in both systems; cytoplasm aggregated in cells adjacent to cells under fungal attack, but not adjacent to needle-induced wound plugs. Hence both non-specific and specific mechanical interactions are involved in the host response to the pathogen.

Hyphal growth directions

The perception of correct direction of hyphal growth, or movement of the pathogen to an appropriate infection site and recognition of that site, is essential for fungal infection (Hoch & Staples, 1991). In rust infection, the fungal spore germinates and produces a germ tube on the surface of the leaf. When the germ tube senses the presence of a stoma – the aperture in the leaf surface enabling gas (including water vapour) exchange – under its tip, the tube stops growing and enlarges to form an appressorium. The tip then balloons out to form an infection peg, which penetrates the leaf and eventually forms a haustorium inside a cell. The rust germ tube must seek and recognise leaf stomata, the only sites that trigger appressorium formation and allow pathogen entry.

Are mechanical effects involved in controlling the direction of movement or growth and in subsequent infection? Answering the questions of how fungal cells sense the direction to grow and the correct site for the infection event provides important knowledge necessary to develop unique and effective disease control strategies. It may be possible to engineer plants so that their surfaces lack proper signals. Currently, there is considerable information about the signals involved in growth orientation and in infection site recognition for some fungi, but we are only beginning to delve into the mechanisms of signal reception and response (Hoch & Staples, 1991).

Some fungal pathogens are not specific for particular infection sites, but others show infection site specificity. Growth is orientated toward these preferred sites by either chemical or physical signals; separating the two is difficult. Typically, hyphae will grow along or at right angles to the anticlinal cell walls on the leaf surface. Wynn (1976) used plastic replicas of leaves and showed that rust fungi would produce appressoria over plastic stomatal impressions exactly as on native leaf stomata; such replicas are free of leaf nutritional or chemical signals, showing that appressorium formation is triggered by topographical features inherent on the leaf surface (Hoch & Staples, 1991).

Dickinson (1977, and earlier papers since the 1940s) showed that spores could germinate and grow hyphae on plastic substrates, and differentiation of appressorium could be induced by scratches or other features of surface topology. More recently, responses of animal cells and spore germ-tubes to defined topographies fabricated by electron-beam lithography have been studied (Clark *et al.*, 1987, 1991; Hoch *et al.*, 1987). When spores of the bean rust were germinated, the hyphae would differentiate to form an appressorium when they crossed either a ridge or a groove with a sharp elevation change of about 0.5 μm (Hoch & Staples, 1991). Ridges or grooves of 1 or 0.25 μm elevation change did not induce appressorium formation. Significantly, the 'lips' around the stomatal guard cells are 0.48 μm high, indicating that site recognition is based solely on thigmotropic sensing by the bean rust; similar effects have not been demonstrated for other species. How does the pathogen discern an elevation change of between 0.2 and 0.5 μm, and initiate an appressorium? As yet, there is little information about the mechanisms, but it is likely that they parallel the mammalian systems perhaps involving actin cable distortion (see review by Dunn, 1991), membrane deformation and calcium or cyclic AMP signal pathways. The fungal system may be amenable to genetic analysis both by production of sensing mutants and by genetic transformation with sense and antisense genes, and hence advance our understanding of the molecular biology of mechanical sensing and interactions.

Non-host resistance and mechanical effects

Most plant species, and indeed animals, are uniformly and completely resistant to most pathogens – they are non-hosts to the pathogen, showing 'non-host resistance' (Niks & Rubiales, 1993). Although a self-evident result, it is nevertheless worth investigating why tomato is a non-host to most pathogens of cereals; in particular, can the resistance mechanisms be transferred or used in different crop species? In the short term, tomato is unlikely to be a useful source of resistance genes to incorporate in cereals, but there are non-hosts to many important crop pathogens in closely related wild species, where gene transfer by crossing and backcrossing is likely to be possible. *Hordeum chilense* is a wild South American species of barley which shows non-host resistance to wheat pathogens such as rust (Rubiales & Niks, 1994). The spores germinate on the surface of the leaves, but the infection tubes overgrow the stomata and very few appressoria are formed. The low appressorium formation frequency was considered to be due to poor stoma recognition – a mechanical effect – rather than poor spore germination or initial growth (Niks, 1986; Rubiales & Nicks, 1994).

Hybrids between *Hordeum chilense* and wheat can be made. After backcrossing, lines with small chromosome segments from *H. chilense* can be generated, and some disease resistances may be expressed in the alien recombinant line (Anamthawat-Jónsson *et al.*, 1990). Unfortunately, the F_1 *H. chilense* × wheat hybrid does not show rust resistance, so the *H. chilense* genes are presumably unexpressed. It is possible that this is an effect related to the architecture of the nucleus (Heslop-Harrison, 1990), and that manipulation of chromosome position may change the dominance of some wheat genes and hence enable expression of the *H. chilense*-origin genes.

Cell–cell interactions during reproduction

Plant reproduction

The cells association with reproduction in angiosperm plants are involved in many mechanical intercellular interactions which control growth, gene expression and even cell death. During plant reproduction, the pollen grain produces a tube that grows at its tip, through the intercellular spaces in the style, to deliver the gametes to the egg and central cell nuclei of the female embryo sac. The pollen first hydrates on the stigma surface, and germinates to produce a tube which penetrates the stigma, and then follows the pollen tube transmitting pathway. The pollen tube pathway is demarcated by both chemical and morphological adaptations. Different species of plants have different morphological adaptations for both the stigma surface to which the pollen attaches and to the transmitting tract. Stigma surfaces range from wet to dry (Heslop-Harrison, 1981, 1992a) and the transmitting tracts vary from being demarcated with special cell types to those in some stigmata which are hollow.

The interactions of the pollen tube with the stigma are interesting scientifically and important economically: failure of fertilisation regularly leads to crop failure. Wide hybrids between different species are required to increase genetic variation in crops (such as those discussed above, p. 113; Heslop-Harrison, 1992b), but cannot be made because the interactions fail (Heslop-Harrison & Schwarzacher, 1993). What is the nature of signal transduction and first responses of pollen tubes to mechanical stimuli, and how are mechanical and other stimuli related? How is the tube guided over the surface of the ovule to the embryo sac? The interaction between the two independent organisms in the pollen–stigma system has only been studied peripherally from the view of cellular mechanical interactions.

In vitro experiments

Because pollen of many species may germinate and the tubes grow independently in simple liquid or semi-solid media, pollen tube growth and development can be studied *in vivo*, when the tube is not interacting with cells of another plant. Pollen grown *in vitro* on the surface of a semi-solid medium shows gentle nutational – slow wandering – movements, as well as sharper changes in direction (Heslop-Harrison, 1987). The interactions that occur when one tube grows into another lying across its path depend on the angle of meeting. If low, the second tube will turn to grow parallel to and along the first tube, and then grow away from it, probably by chance nutational movements. If the tubes meet at a higher angle (perhaps 75° or more), the second tube may grow over the first with little change in growth direction (J. S. Heslop-Harrison, unpublished data). As yet, plastic replicas of styles have not been used to examine pollen growth, but their use should show which growth responses are mechanically triggered. Knowledge of mechanical interactions which occur during tube growth may assist with obtaining *in vitro* fertilisation or characterising hybrid combinations where mechanical guidance failures prevent the tube finding the egg apparatus.

Pollen tubes and stigma guidance

Studying interactions between pollen tubes and the style is difficult, because they involve mechanical, chemotropic, electrical and nutritional interactions that are difficult to isolate from each other. There is a close co-adaptation between the pollen grain and pollen tube size, and the stigma – presumably a co-evolution of sizes enabling correct interactions. What are the cues given to the pollen about the stigma size, and how does the pollen receive these signals? How specific are the cues?

Many models have been described to show how the pollen tube grows towards and finds the egg apparatus (see Heslop-Harrison, 1987). The idea that the tube simply locates a target is not correct – the ovule may be more than 300 mm away from the site of pollen grain capture in maize (*Zea mays*), and the tube must grow this length to deliver the sperm cells. It is hard to imagine that the signalling involves long-range signals or gradients. It is more likely that directionality is initially imposed on the pollen tube by stigma morphology, and then the tube is constrained to grow in a particular way and direction – control by the architecture of the pollen tube pathway and the properties of the cell that define it. The mechanical and chemotropic guidance mechanisms implied by a targeting hypothesis would have very different specificities from those found in both normal pollen–stigma interactions and interactions between a stigma and pollen from another species or genus of plant.

Stigma morphology generally ensures that, when the tube enters the transmitting tract, it is unlikely to leave it (Heslop-Harrison & Heslop-Harrison, 1981; Heslop-Harrison, Heslop-Harrison & Reger, 1985). For example, in *Zea mays*, stigma trichomes guide tubes towards the stigma axis as they emerge from the pollen grain, and the tube penetrates the stigma pellicle. Once at the base of the trichome, the tube is generally constrained to grow in a single direction by the cell wall structures. In fact, different tubes can grow stably in both directions through the tract, which can be shown either by grafting inverted sections of style into the stigma, when the tube will still grow straight in the same direction (Iwanami, 1959), or by pollinating cut ends of the style. The study of exceptional interactions also indicates the importance of the relative sizes of stigma cells and pollen tube. If maize is pollinated with pearl millet, *Pennisetum typhoides* (syn. *P. americanum*), pollen, the pollen will germinate and grow well. However, the tube, which is smaller than that of maize, is not guided correctly and grows in both directions in the style, even reversing direction.

Another error in pollen tube guidance was highlighted by analysis of the behaviour of alien pollen tubes when they reach the ovary in wheat. The defined nature of the barrier was not suspected from examination of pollen tubes in intraspecific crosses, since tubes grow through the zone without any check to growth. However, Sitch & Snape (1987) examined wheats with different genotypes at two mapped crossability loci. *Hordeum bulbosum* pollen germinated and tube growth as far as the base of the stigma was similar in all genotypes. The frequency of pollen tube penetration of the ovary wall was severely reduced when the dominant *Kr* alleles were present, showing that a genetically controlled barrier operated at this point, and perhaps it involves mechanical stimuli to the pollen tube tip.

We have investigated the ion distribution at the base of the stigma in *Zea mays*, and found very high levels of calcium in the zone between the ovule wall and nucellus where the pollen tube grows (Heslop-Harrison & Reger, 1986). Perhaps a tendency to grow in or towards high calcium concentrations, combined with mechanical effects, enables *Z. mays* pollen grains to follow the nucellar surface to the micropyle. Artificial pollination of a sectioned *Z. mays* ovary wall and nucellus with *P. americanum* shows that the pollen grains germinate and, after some random growth, can follow the normal pollen tube growth pathway for some distance. Further work will help establish whether the stimulus is mechanical, nutritional, chemotactic or chemotropic. No successful fertilisations have yet been achieved by this semi-*vitro* method, and the pollen tube usually bursts after several hours' growth. However, it would be important if systems can be developed which will increase the types of wide hybrids that can be produced sexually.

Conclusions

The data reviewed above indicate current work on some selected mechanical cell interactions which occur in plants. Many other mechanical interactions have been studied in plants, including thigmotropic responses in genera such as *Arabidopsis, Dionea* and *Mimosa*, differentiation in response to mechanical damage or contact, including cork production and grafting, responses to gravity, or situations where cells are crushed by growth of other cells. Parasitic and hemiparasitic plants have many important interactions with their hosts; many beneficial associations between plants and fungi or bacteria (including mycorrhizeae and nitrogen-fixing organisms) may also involve mechanical interactions between two organisms.

While much of the work on mechanical cell interactions is still descriptive, increasingly a genetical and molecular approach is being used to study cell behaviour. Some responses involve reorientation of growth or movement, and can be achieved by changes in the configuration of the cytoskeleton – including at least actin and tubulin components. Other responses involve changes in gene activity, and it will be important to find whether there is any physical reorganisation of the spatial positioning of genes within the nucleus in response to changing transcription and processing demands.

Many of the results discussed have direct relevance and consequences for practical problems – those in disease resistance and the production of wide hybrids have been emphasised. Some have direct parallels with other topics discussed in this volume; the comparison of perception, responses and transduction mechanisms in plant and animal cells may enable isolation and characterisation of the universal components of phenomena related to the mechanical interactions and the responses of cells.

References

Aist, J. R. (1976). Papillae and related wound plugs of plant cells. *Annual Review of Phytopathology* **14**, 145–63.

Anamthawat-Jónsson, K., Schwarzacher, T., Leitch, A. R., Bennett, M. D. & Heslop-Harrison, J. S. (1990). Discrimination between closely related Triticeae species using genomic DNA as a probe. *Theoretical and Applied Genetics* **79**, 721–8.

Bartholdi, M. F. (1991). Nuclear distribution of centromeres during the cell cycle of human fibroblasts. *Journal of Cell Science* **99**, 255–63.

Borden, J. & Manuelidis, L. (1988). Movement of the X chromosome in epilepsy. *Science* **242**, 1687–91.

Carter, K. C., Bowman, D., Carrington, W., Fogarty, K., McNeil, J. A., Fay, F. S. & Lawrence, J. B. (1993). A three-dimensional view of precursor messenger RNA metabolism within the mammalian nucleus. *Science* **259**, 1330–5.

Clark, P., Connolly, P., Curtis, A. S. G., Dow, J. A. T. & Wilkinson, C. D. W. (1987). Topographical control of cell behaviour I. Simple step cues. *Development* **99**, 439–48.

Clark, P., Connolly, P., Curtis, A. S. G., Dow, J. A. T. & Wilkinson, C. D. W. (1991). Cell guidance by ultrafine topography *in vitro*. *Journal of Cell Science* **99**, 73–7.

Cole, T. & Hoch, C. E. (ed.) (1991). *The Fungal Spore and Disease Initiation in Plants and Animals*. New York: Plenum Press.

Darvill, A. G., Albersheim, P., Bucheli, P. *et al.* (1989). Oligonucleotides: plant regulatory molecules. In *NATO Advanced Workshop on Molecular Signals in Microbe–Plant Symbiotic and Pathogenic Systems*, ed. B. J. J. Lugtenberg, pp. 41–8. New York: Springer.

Dickinson, S. (1977). Studies in the physiology of obligate parasitism X. Induction of responses to a thigmotropic stimulus. *Zeitschrift für Phytopathologie* **89**, 97–115.

Dunn, G. A. (1991). How do cells respond to ultrafine surface contours? *Bioessays* **13**, 541–3.

Heslop-Harrison, J. (1987). Pollen germination and pollen-tube growth. *International Review of Cytology* **107**, 1–78.

Heslop-Harrison, J. & Heslop-Harrison, Y. (1981). The pollen–stigma interaction in the grasses. 2. Pollen-tube penetration and the stigma response in *Secale*. *Acta Botanica Neerlandica* **30**, 289–307.

Heslop-Harrison, J. S. (1990). Gene expression and parental dominance in hybrid plants. *Development* (Suppl.), 21–8.

Heslop-Harrison, J. S. (1992a). The angiosperm stigma. In *Sexual Plant Reproduction*, ed. M. Cresti & A. Tiezzi, pp. 59–68. Heidelberg: Springer.

Heslop-Harrison, J. S. (1992b). Natural and artificial hybrids in the grasses. In *Angiosperm Pollen and Ovules*, ed. D. Mulcahy & G. Bergamini-Mulcahy, pp. 408–13. New York: Springer.

Heslop-Harrison, J. S. & Reger, B. J. (1986). X-ray microprobe mapping of certain elements in the ovary of *Zea mays* L. *Annals of Botany* **57** 819–22.

Heslop-Harrison, J. S. & Schwarzacher, T. (1993). Molecular cytogenetics – biology and applications in plant breeding. *Chromosomes Today* **11**, 191–8.

Heslop-Harrison, Y. (1981). Stigma characteristics and angiosperm taxonomy. *Nordic Journal of Botany* 1, 401–20.

Heslop-Harrison, Y., Heslop-Harrison, J. & Reger, B. J. (1985). The pollen–stigma interaction in the grasses. 7. Pollen-tube guidance and the regulation of tube number in *Zea mays* L. *Acta Botanica Neerlandica* 34, 193–211.

Hoch, H. C., Staples, R. C. Whitehead, B., Coeau, J. & Wolf, E. D. (1987). Signaling for growth orientation and cell differentiation by surface topography in *Uromyces*. *Science* 235, 1659–62.

Hoch, H. C. & Staples, R. C. (1991). Signaling for infection structure formation in fungi. In *The Fungal Spore and Disease Initiation in Plants and Animals*, ed. T. Cole & C. Hoch, pp. 25–46. New York: Plenum Press.

Iwanami, Y. (1959). Physiological studies of pollen. *Journal of Yokohama Municipal University* 116, 1–137.

Jefferson, R. A., Kavanagh, T. A. & Bevan, M. W. (1987). GUS fusions: β-glucuronidase as a sensitive and versatile gene fusion marker in higher plants. *EMBO Journal* 6, 3901–7.

Lawrence, J. B., Singer, R. H. & Marselle, L. M. (1989). Highly localized tracks of specific transcripts within interphase nuclei visualized by *in situ* hybridization. *Cell* 57, 493–502.

Lawrence, J. B. & Singer, R. H. (1985). Quantitative analysis of *in situ* hybridization methods for the detection of actin gene expression. *Nucleic Acids Research*, 1777–99.

Linskens, H. F. & Heslop-Harrison, J. (ed.) (1984). *Cellular Interactions. Encyclopaedia of Plant Physiology*, New Ser., Vol. 17, ed. A. Pirson & M. H. Zimmermann. New York: Springer-Verlag.

Matzke, M. A. & Matzke, A. J. M. (1990). Gene interactions and epigenetic variation in transgenic plants. *Developmental Genetics* 11, 214–23.

Matzke, M. A. & Matzke, A. J. M. (1991). Differential inactivation and methylation of a transgene in plants by two suppressor loci containing homologous sequences. *Plant Molecular Biology* 16, 821–30.

Nichols, S. P. (1925). The effect of wounds upon the rotation of the protoplasm in the internodes of *Nitella*. *Bulletin of the Torrey Botanical Club* 52, 351–63.

Nicholson, R. L. & Epstein, L. (1991). Adhesion of fungi to the plant surface: prerequisite for pathogenesis. In *The Fungal Spore and Disease Initiation in Plants and Animals*, ed. T. Cole & C. Hoch, pp. 3–23. New York: Plenum Press.

Niks, R. E. (1986). Failure of haustorial development as a factor in slow growth and development of *Puccinia hordei* in partially resistant barley seedlings. *Physiological and Molecular Plant Pathology* 28, 309–22.

Niks, R. E. & Rubiales, D. (1994). Use of non-host resistance in wheat breeding. In *Biodiversity and Wheat Improvement*, ed. A. Damania. London: Wiley and Sons (in press).

Nims, R. C., Halliwell, R. S. & Rosberg, D. W. (1967). Wound healing in cultured tobacco cells following microinjection. *Protoplasma* **64**, 305–14.

Rubiales, D. & Niks, R. E. (1992). Low appressorium formation by rust fungi on *Hordeum chilense* lines. *Phytopathology* **82**, 1007–12.

Russo, V. M. & Bushnell, W. R. (1989). Responses of barley cells to puncture by microneedles and to attempted penetration by *Erysiphe graminis* f. sp. *hordei*. *Canadian Journal of Botany* **67**, 2912–21.

Shirras, A. D. & Northcote, D. H. (1984). Molecular cloning and characterisation of cDNAs complementary to mRNAs from wounded potato (*Solanum tuberosum*) tuber tissue. *Planta* **162**, 353–60.

Sitch, L. A. & Snape, J. W. (1987). Factors affecting haploid production in wheat using the *Hordeum bulbosum* system. 1. Genotypic and environmental effects on pollen grain germination and the frequency of fertilization. *Euphytica* **36**, 483–96.

Stanford, A., Bevan, M. & Northcote, D. (1989). Differential expression within a family of novel wound-induced genes in potato. *Molecular and General Genetics* **215**, 200–8.

Stanford, A. C., Northcote, D. H. & Bevan, M. W. (1990). Spatial and temporal patterns of transcription of a wound-induced gene in potato. *EMBO Journal* **9**, 593–603.

Sussex, I., Ellingboe, A., Crouch, M. & Malmberg, R. (1985). *Current Communications in Molecular Biology: Plant Cell/Cell Interactions*. New York: Cold Spring Harbor Laboratory.

Wynn, W. K. (1976). Appressorium formation over stomates by the bean rust fungus: response to a surface contact stimulus. *Phytopathology* **66**, 136–46.

Xing, Y., Johnson, C. V., Dobner, P. R. & Lawrence, J. B. (1993). Higher level organization of individual gene transcription and RNA splicing. *Science* **259**, 1326–30.

A. S. G. CURTIS

Mechanical tensing of cells and chromosome arrangement

Many reports have been made describing how the application of mechanical forces to cells leads to changes in gene expression. Several of the contributions to this volume report on that matter. In addition it is pertinent to quote papers by Goldspink and his colleagues (see, for example, Goldspink *et al.*, 1992). The next question that arises is: how do such force applications lead to changes in gene expression?

Amongst the various theories that have been proposed to explain these effects on gene expression are the following.

1 That mechanical tension opens specific or non-specific ion channels that lead to changes in the cell, either operating through signalling systems or simply facilitating the entry of nutritive factors into the cell. Since stretch receptors are well known in neurones (e.g. Blackshaw & Thompson, 1988; Sigurdson & Morris, 1989) this theory has direct experimental support as a possible mechanism, although the activation of stretch reception does not in itself lead to transcriptional changes.

2 That mechanical tension affects the cytoskeleton and that this in turn operates either translationally in the cytoplasm or on nuclear events to alter expression (Pender & McCulloch, 1991).

3 That mechanical tension activates signalling processes at the plasma membrane (Horoyan *et al.*, 1991). This suggestion is explored by several other contributors to this volume, see for example the chapter by D. Jones.

4 That all that mechanical tension does, whether acting directly on the cells or more indirectly, is to provide more stirring of the medium around the cells or better access of metabolites to the cells and better removal of waste products.

Society for Experimental Biology Seminar Series 54: *Biomechanics and Cells*, ed. F. Lyall & A. J. El Haj. © Cambridge University Press 1994, pp. 121–130.

Dunn & Ireland (1984) suggested that this accounted for the effects of mechanical tension on cell growth in fibroblasts.

The second type of explanation can be related to the suggestion by Bennett (1987), Heslop-Harrison *et al.* (1991) and Leitch *et al.* (1991) that chromosomes in interphase nuclei have defined positions. Heslop-Harrison produced much evidence, by laborious computer reconstruction of chromosomal position with probes, that chromosomes have defined if approximate positions in interphase nuclei of plants and of a human cell line. If this theory is correct it is easy to proceed one step further and suggest that the application of mechanical tension to the nuclei in one direction might alter these arrangements. For example, the tension might orient parts of chromosomes and make access of ribonucleotides to the transcription site more easy or more difficult. Tensional forces might affect the accessibility of certain parts of the chromosome to reaction with various proteins. Finally, tensional forces might simply affect such matters as base-pairing during transcription by altering the general shape of the DNA. More complex models for expression control by tension might be considered.

This short contribution describes an attempt to test the hypothesis. Before describing these experiments it is relevant to comment that a literature search has failed to find any previous work directly relevant to this question published in the past 4 years. It is also pertinent to remark that should evidence be found for the mechanism it is quite possible that the application of mechanical tension to the cell operates through the cytoskeleton and then the nucleoskeleton to affect the chromosomes. There are also a number of prerequisites that must be met in this type of experiment. These are described in the next section.

Before proceeding to the next section it is worth noting that in life mechanical tensions, both transient and persistent, are inflicted on cells. An extreme example is provided by the cells of the vocal cords of a singer *in fortissimo*. However, despite this remarkable case and some others like it, few cells are probably subjected to more than a 10% tensional extension in length over a short period. We have, however, to exclude the multinucleate muscle cells from this statement.

The ideal experiment to test whether tension affects chromosome packing

There are a number of important considerations.

1. The way in which the mechanical tension is applied, and its magnitude, must be described

Cells are known to react to the space available for spreading and to contact with other cells by changes in transit through the cell cycle and by changes in cell shape. Changes in cell shape might in themselves be enough to permit changes in permeation of nutrients, etc. into or out of the cell. It was for this reason that Curtis & Seehar (1978) applied mechanical tension as a series of stretching and relaxations to a cell sheet grown on a mesh. Though this might be sufficient to overcome problems about changes in the area of the cells it has the shortcoming that the attachment of the cells to each other could lead to modification of the pattern of mechanical stresses in these sheets in ways which might differ from place to place. Thus a thorough-going analysis of the effects of tension should be carried out on isolated cells grown on an appropriate substrate. Obviously the mesh type of support used earlier is inappropriate. Elastic substrates might be more suitable because cells can be grown in isolation on them and the mechanical tensing can be cyclical so that no long-term change in cell shape or area occurs. Banes *et al.* (1985) introduced an ingenious method of applying tension repetitively. Cells grown on an elastic sheet were subjected to stretching and to compression by flexing the sheet by a vacuum device (Fig. 1). This device stretches the cells to different extents depending whereabouts they are on the deformable surface. It also has the problem that stretching occurs in at least two dimensions while the cells are compressed in the third dimension. Another type of device is a simple elastic sheet which is stretched in one dimension. This leads inevitably to some compression in the other dimension unless the sheet is attached to a rigid frame on the other two sides. Thus three types of device (see Fig. 1) provide mechanical stress in three, two or one dimension(s). It would be interesting to compare the results of the three different approaches.

2. Establishing possible chromosomal displacements in interphase nuclei

The basic approach here is to use *in situ* hybridisation of DNA probes to chromosomal DNA to identify specific regions of one or more chromosomes. Because of the difficulties with observation of more than three different probes which arise from the limited number of fluorescent probe systems operating at different excitation and emission wavelengths, we are in practice limited to detecting at best two or three probe sites in any one set of experiments. With the exception of probe binding sites on X and Y chromosomes this means that we normally should obtain paired probing sites for each specific probe. There is of course the unfortunate possibility that they lie on top of each other from the

124   A. CURTIS

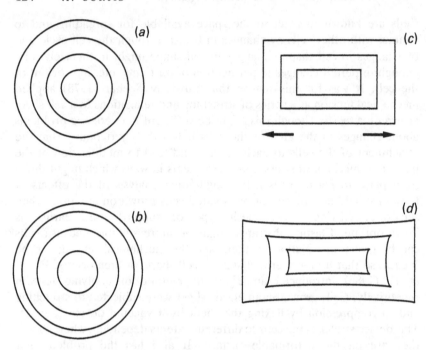

Fig. 1. Types of elastic substrate or cell sheet distortion that may occur in various experimental methods for stretching cells, showing non-linear effects. (a) Circular areas with equidistant contours which are distorted in 'blister' fashion to produce (b). Note in (b) (after stretching) that some contours have widened apart and others closed in. Stretching is very anisotropic. In (c) a rectangular area is gripped at right and left edges and tension applied in the directions of the arrows. Note the types of distortion produced in (d).

viewpoint of microscopy but it is clear that in this case, as for other arrangements, confocal scanning microscopy will be the method of choice for investigation of probe positions.

How do we discover whether application of mechanical forces affects interphase chromosome displacement? Obviously we need to make measurements of the extent of separation of various parts of given chromosomes from each other. Ideally the measurement would be of the relative positions of two or three sites in regions on chromosomes, which are already known to contain genes that respond to mechanical stimulation. Thus any change in their relative positions would indicate some degree of rearrangement. With existing methods the accuracy of such measurements would be no better than 0.5 μm. There is probably no point in relating these measurements to other features of the cell such as nucleolar position or the centre of gravity of the cell since there is no evidence that either of these features has any clear relation to chromosome position. There may be some point in relating such measurements to the axis of stretch.

I have adopted a simpler approach in initial work. I have used centromeric probes, in this case for human chromosome 12. If mechanical forces produce a general reorganisation of the chromosome pattern in the nucleus it might be expected that the positions or variance in position of the centromere of two pairs would change. Figure 2 illustrates some possibilities.

If stretching really affects accessibility of relatively large molecules to DNA sequences then in principle the intensity of staining with various sized probes or antibodies might provide a measure of this, though it should be borne in mind that the permeabilisation or pre-hybridisation treatments might grossly alter accessibility.

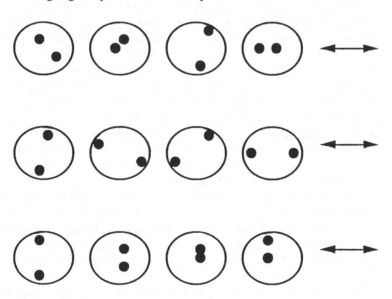

Fig. 2. Possible positions of paired centromeres in the nucleus. *Top row*, four nuclei with random arrangements of centromeres with no relationship to the direction of stretch (see arrows) and wide variance in their separation. *Middle row*, centromeres responding by increasing their spacing with reduced variance in spacing. *Bottom row*, centromeres reacting by lining up so that the axis joining pairs is at 90° to the direction of stretch shown by the arrows.

3. What is the appropriate regime of mechanical force application?

Until a thorough investigation of the effects of magnitude, style of application, duration and frequency of application of mechanical stress to a cell have been carried out it is impossible to state what style of application will yield the most interesting results. In situations where heartbeat or repetitive muscular action are likely to provide *in vivo* a dominant style of application of forces it is obviously sensible to investigate situations which mimic the natural application. Curtis & Seehar (1978) found that low frequencies (*c.* 1 Hz) of stress application produced effects on cell cycle rate. In the work described here I have chosen to study a simple rapid extension of the cell in one dimension. This may not correspond to any particular biological event but is relatively simple in nature.

Initial experimental results

HGTFN cells, a human capillary endothelial line of cells isolated in this laboratory from granulation tissue in a wounded hand tendon, were used for this investigation. This line of cells has been maintained in my laboratory for 26 passages (at the time of writing) and is recognised as being endothelial by its possession of a low density lipoprotein marker, by having a positive reaction for Factor VIII and by the possession of surface receptors for the lectins from *Ulex europaeus* and from *Griffonia simplicifolia*. The cells were grown in Hams F10 medium with 3% fetal calf serum and an insulin–transferrin–selenite supplement.

Cultures were grown either on glass coverslips or on silicone rubber (Esco, Ltd) at very low density so that the majority of cells were isolated at the time of application of mechanical tension.

In order to tense the cells they were grown on silicone rubber sheets held in the tensing device (see Fig. 3) for 24 h. The sheets were extended by moving the free clamp closer to the stop on the holder. In the majority of experiments the sheet was stretched between 115 and 130% (over about 1 min.). The cells were left growing on the stretched sheet for 3 hours.

The cells were then fixed (paraformaldehyde followed by permeabilisation with Triton X-100) and hybridised with a centromeric probe for chromosome 12 (O1223, insert size 1.35 kb) using the probe and techniques kindly made available to me by Dr Elizabeth Boyd of the Department of Human Genetics, Glasgow University. The probe had been biotinylated so that a fluorescent (Cascade Blue) avidin stain was used to reveal the centromeres. When hybridisation was carried out with low

Fig. 3. Photograph of device for applying mechanical tension to cells. The device lies within a 65 mm internal diameter Petri dish.

stringency a certain amount of non-specific labelling of DNA took place which was useful to show the limits of the nucleus. Phase contrast microscopy could be used to reveal general cell morphology.

The method of analysis used has been discussed and described above. The distance between centromeres was measured with an ocular micrometer. Two centromeres were seen in the majority of nuclei. In some nuclei only one centromere was seen but in such cases the 'dot' was often larger than the ones where two centromeres could be visualised in one cell, and such an appearance is probably due to one centromere overlying the other. The distance in such cells was scored as 0 μm though there

Table 1. *Effect of mechanical tension on separation on centromere 12 pairs in human endothelial cells*

Control cells	Centromeric separation 5.9 μm SD 3.5
Tensed cells	Centromeric separation 9.4 μm SD 5.2

t test, $t = 2.95$; $p = 0.05$; df =57.

must be some distance in the z-axis. Ideally the distance between centromeres should have been measured using confocal scanning microscopy so that z-axis distances as well as X and Y distances could have been measured. In future work it would be very desirable to use a second probe with a distinctive colour.

The 120% stretching of the substrate appeared to be accommodated by the majority of cells without difficulty. A few cells rounded up, possibly due to the distortion of the substrate and loss of adhesion. This needs further investigation. There was no evidence that cells elongated at 90° to the direction of stress application reacted differently from those elongated parallel to the stress.

Reaction of centromeres to stretch

There was no obvious orientation of the centromeres in relation to the direction of stretch. Consequently in this preliminary study no statistical search was made for evidence of such an effect. There was a clear increase in centromere separation (see Table 1) which is significant at $p < 0.01$. There was no evidence that the variance of centromere separation increased testing with the z test. Typical views of such hybridisations are shown in Fig. 4. Preliminary experiments suggest that similar effects may develop within 20 min of stretching.

Conclusions

This preliminary investigation shows that it is possible that application of a mechanical force stretching a cell by c. 120% can lead to increased centromere separation on chromosome 12. Further investigation is clearly required as to the effects of similar stretch on the relative positions of other chromosomal markers, especially those associated with known areas of increased gene expression in response to mechanical stress.

Curtis & Seehar (1978) have suggested that growth control in confluent cultures might be affected by mechanical tension. This suggestion was at variance with the commonly held concept at that period that growth control was a diffusion limited process. The evidence for this came from

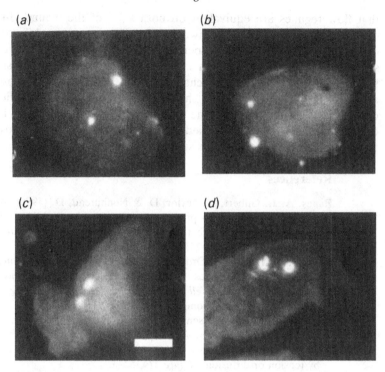

Fig. 4. Centromere 12 hybridisations in mechanically stressed (*a*, *b*) and control (*c*, *d*) HGTFN cells, photographed with a NA 0.9 objective. Note the greater variability in centromere position in the stressed set *a* and *b*. Scale bar = 5 μm.

the use of the Riddle pump in which cultures with actively recirculated medium grew to higher cell density than those in still medium. Curtis & Seehar suggested that mechanical explanations might be more accurate since they had taken much care to minimise stirring effects. Dunn & Ireland (1984) carried out an ingenious experiment which appears to restore belief in the diffusion limitation theory. In this they rotated a disk close to the surface of the culture. A wound line in the culture normally allows increased DNA synthesis in the cells close to the wound – arguably because diffusion processes are less limited when the cells are at an edge. In their rotation experiment cells only showed increased DNA synthesis in the downstream edge of the wound, where the cells were exposed to fresh medium, and not in the upstream edge. Nevertheless this interpretation of the experiment depends on the presumption

130   A. CURTIS

that flow regimes are equivalent on both sides of the wound. In fact on the upstream side the cells would experience tension while on the downstream side they should experience a mild compression. In the present experiments the cells were in isolation so that diffusion effects should be equal on all cells. But mechanical stress is altering chromosomal arrangement. It remains to be shown unequivocally that alteration of chromosomal position alters gene transcription, and that the cells in the culture have transcription modified by stress, but we are now in a position to test it and preliminary results support the idea.

References

Banes, A. J., Gilbert, J., Taylor, D. & Nonbureau, D. (1985). A new vacuum-operated stress-providing instrument that applies static or variable duration cyclic tension or compression to cells *in vitro*. *Journal of Cell Science* **75**, 35–42.

Bennett, M. D. (1987). Ordered disposition of parental genomes and individual chromosomes in reconstructed plant nuclei and their implications. *Somatic Cell and Molecular Genetics* **13**, 463–6.

Blackshaw, S. E. & Thompson, S. W. N. (1988). Hyperpolarizing responses to stretch in sensory neurones innervating leech body wall muscle. *Journal of Physiology* **396**, 121–37.

Curtis, A. S. G. & Seehar, G. M. (1978). The control of cell division by tension or diffusion. *Nature* **274**, 52–3.

Dunn, G. A. & Ireland, G. W. (1984). New evidence that growth in 3T3 cell cultures is a diffusion limited process. *Nature* **312**, 63–5.

Goldspink, G., Scott, A., Loughna, P. T., Wells, D. J., Jaenicke, T. & Gerlach, G. F. (1992). Gene expression in skeletal muscle in response to stretch and force generation. *American Journal of Physiology* **262**, 199.

Heslop-Harrison, J. S., Mosgoller, W., Schwarzacher, T. & Leitch, A. R. (1991). Volumes and positions of chromosomes in reconstructions of fibroblasts. *American Journal of Human Genetics* **49**, 380.

Horoyan, M., Benoliel, A.-M., Capo, C. & Bongrand, P. (1991). Localization of calcium and microfilament changes in mechanically stressed cells. *Cell Biophysics* **17**, 243–56.

Leitch, A. R., Schwarzacher, T. Mosgoller, W., Bennett, M. D. & Heslop-Harrison, J. S. (1991). Parental genomes are separated throughout the cell-cycle in a plant hybrid. *Chromosoma* **101**, 206–13.

Pender, N. & McCulloch, C. A. G. (1991). Quantitation of actin polymerization in two human fibroblast sub-types responding to mechanical stretching. *Journal of Cell Science* **100**, 187–93.

Sigurdson, W. J. & Morris, C. E. (1989). Stretch-activated ion channels in growth cones of snail neurons. *Journal of Neuroscience* **9**, 2801–8.

B. REIPERT

Alterations in gene expression induced by low-frequency, low-intensity electromagnetic fields

The comprehensive electrification of modern societies has created conditions where living organisms are exposed to 50 or 60 Hz power frequency electric and magnetic fields which are, although usually weak, clearly considerably above the naturally occurring ambient fields of about 10^{-4} V m^{-1} (electric field) and up to 10^{-4} T (magnetic fields). To demonstrate the size of the fields concerned, Table 1 summarises both the field strengths of electric fields and the flux densities of magnetic fields that are found near power lines (NRPB, 1992a). However, both electric and magnetic fields are encountered wherever electric current is either produced or distributed. To illustrate the situation, a number of examples of magnetic flux densities measured at various distances from commonly used domestic appliances are summarised in Table 2 (NRPB, 1992b). The electric field strengths close to domestic appliances are in the range of about 200–300 V m^{-1}.

A series of epidemiological studies including residents who live close to high voltage power lines has shown a possible link between the exposure to weak electromagnetic fields (EMF) and an increased risk for the development of leukaemia, in particular leukaemia in children (Wertheimer & Leeper, 1979; Savitz et al., 1988; Coleman et al., 1989; London et al., 1991; Feychting & Ahlbom, 1992). However, the calculated risk factor obtained from those studies is quite small (in the range of about 1.5–3.0). In general, however, the whole spectrum of epidemiological studies dealing with the exposure to EMF seems to obtain rather contradictory results. Nonetheless, public awareness of a potential hazard has been raised and a considerable number of experimental studies stimulated over the last ten years aimed at determining whether or not weak EMF might indeed be able to interact with biological systems and cause detectable alterations in cellular functions. In spite of these studies, however, no clear consensus has emerged as to whether such EMF express

Society for Experimental Biology Seminar Series 54: *Biomechanics and Cells*, ed. F. Lyall & A. J. El Haj. © Cambridge University Press 1994, pp. 131–143.

Table 1. *Magnetic flux densities and electric field strengths near power lines*

	Maximum at ground level beneath power lines	25 m lateral displacement from centreline
Magnetic flux density (μT) near power lines of:		
400 kV	40	8
275 kV	22	4
132 kV	10	2
Electric field strength (V m^{-1}) near power lines of:		
400 kV	11 000	1000
275 kV	6 000	200
132 kV	2 000	50

From NRPB (1992a).

Table 2. *Magnetic flux densities at various distances from examples of household appliances*

Appliance	Magnetic flux density (μT) at a distance of:		
	3 cm	30 cm	1 m
Can openers	1–2×10^3	3.5–30	0.07–1
Dishwashers	3.5–20	0.6–3	0.07–0.3
Fluorescence desk lamps	40–400	0.5–2	0.02–0.025
Food mixers	60–700	0.6–10	0.02–0.025
Hair dryers	6–2000	< 0.01–7	< 0.01–0.03
Irons	8–30	0.12–0.3	0.01–0.025
Refrigerators	0.5–1.7	0.01–0.025	< 0.01
Shavers	15–1500	0.08–9	< 0.01–0.3
Television sets	2.5–50	0.04–2	< 0.01–0.15
Toasters	7–18	0.06–0.7	< 0.01
Vacuum cleaners	200–800	2–20	0.13–2

From NRPB (1992b).

biological activity or not. Although there have been many reports in the literature appearing to provide some evidence for biological effects of weak EMF, such effects have not convincingly been established through replication in independent laboratories. In part, this may be due to the uniqueness of some of the experimental protocols adopted in different laboratories, including various exposure parameters (often poorly or inadequately characterised), the particular cell systems used and the endpoints employed to assess possible EMF-induced effects.

The first of these undoubtedly relates to the considerable complexity of attaining appropriate experimental conditions for EMF experiments. The EMF generated can be characterised well by its power density (S) or Poynting vector, the vector product of magnetic field strength (H) and electric field strength (E). The unit of power density is the watt per square metre ($W\ m^{-2}$), the unit of electric field strength is the volt per metre ($V\ m^{-1}$) and the unit of magnetic field strength is the ampère per metre ($A\ m^{-1}$). Frequently, the magnetic flux density (B), with the unit of tesla (T) or Gauss (G) $10^4\ G = 1\ T$ is given instead of the magnetic field strength (H). These are related to each other by the magnetic permeability (μ, $B = \mu H$) of the medium which is expressed in the unit of henry per metre ($H\ m^{-1}$). For most biological materials μ has the free-space value of $4 \times 10^{-7}\ H\ m^{-1}$. Many studies do not include adequate characterization of the EMF generated, making it difficult to compare results obtained in different laboratories. Apart from the actual EMF generated, there is a series of other factors which may interact with each other during the experiment to cause the differences in biological effects (or lack of them) reported by different laboratories. The most important factors in these studies are probably:

1. the local geomagnetic field;
2. alternating background fields at the experimental site;
3. minor differences in temperature;
4. slightly different components in the media used for cell culture;
5. different cell concentrations.

All these parameters should be well characterised and stated clearly with each observation of possible biological effects of EMF.

Among perhaps the most consistent reports to date are those demonstrating EMF-induced changes in gene expression, detected both at the mRNA and at the protein level. Considering the significance any alteration in gene expression might have for the normal functioning of cells, this review will summarise the main reports on EMF-induced changes in gene expression.

134   B. REIPERT

Goodman, Bassett & Henderson (1983) reported that two different signals, a short pulse repeated at 72 Hz and a pulse train at 15 Hz, each increased gene transcription in dipteran salivary gland cells, compared to unexposed control cells. The effects were assessed by transcription autoradiography and cytological nick translation. In addition, after exposing cells to the electromagnetic signals in the presence of [³H]uridine, RNAs of various size classes were separated by sucrose density-gradient centrifugation. The results indicated that exposure to EMF induced an increase of the order of up to 13-fold in precursor uptake into RNAs of the size class consistent with that expected for mRNAs. In subsequent studies, both asymmetric pulsed electromagnetic fields (PEMF) and symmetrical, continuous-wave fields were used (Goodman, Abbot & Henderson, 1987; Goodman & Henderson, 1988; Goodman et al., 1989, 1992a,b; Goodman, 1991). The most dramatic effects were seen with symmetrical, continuous-wave fields, not only in dipteran salivary gland cells but also in the human promyelocytic leukaemic cell line HL-60, which was included in most of their following investigations. In HL-60 cells exposed to EMF, Goodman and co-workers described an increase in the amount of specific mRNAs for histone H2B, β-actin, β-tubulin, v-src and c-myc of the order of up to four-fold compared with control levels. Genes not usually expressed in HL-60 cells, such as α-globin, seemed to be unaffected by EMF. Their hybridisation analyses employed dot-blots for quantification of transcripts by direct counting of radioactive spots. Northern blots were used to confirm the sizes of transcripts. Increased transcript levels were reported after as little as 4 min exposure. Peak levels were seen after different time periods, depending on the actual exposure conditions, but very often after 20 min. Increased transcript levels were sustained for about 2 h. Alterations in transcript levels were also reflected by changes in overall protein synthesis patterns following exposure of both dipteran and HL-60 cells to 60 Hz fields. Clear differences were noted between the response of cells to EMF and the response to stress factors such as heat shock.

Perhaps the most intriguing findings of this group were field strength as well as frequency and time-windows for the maximal biological action of EMF observed (Goodman, 1991). Accordingly, when cells were exposed to magnetic fields with four different amplitudes – 0.57, 5.7, 57 and 570 μT – all rms, corresponding to induced electric fields of the order of 1, 10, 10 and 1000 μV m⁻¹ (Blank et al., 1992b) – the most pronounced changes in the quantities of specific mRNAs were noted at 57 μT whereas the least significant were seen at 570 μT. Similarly, when testing a range of different frequencies from 15 to 150 Hz, the biologically most effective fields were seen at 45 Hz. It should also be noted that, in more recent work, a second such frequency window has been found, at 4.4 kHz, when testing a range of frequencies from 0.072 to 1000 kHz (Wei, Goodman & Henderson, 1990; Goodman, Wei & Henderson, 1991).

These thought-provoking results have stimulated extensive activity in other laboratories. However, the exposure conditions described by Goodman and co-workers were rarely duplicated, and this makes it extremely difficult to compare the various results obtained in the different laboratories. Nevertheless, in several cases similar systems *were* used and comparable results obtained. On the other hand, however, there has also been a series of replication attempts, in which *no* evidence for such effects was elicited (see below).

Phillips & McChesney (1991) reported a number of effects of exposing cells of the human T-lymphoblastoid cell line, CCRF-CEM, to a pulsed magnetic field at 72 Hz. They noted a time-dependent change of precursor uptake into total cellular RNA, mRNA and protein. The changes seen were of the order of 2–3-fold with a peak after 2 h of exposure. In subsequent experiments (Phillips *et al.*, 1992c), they exposed CCRF-CEM cells to 60 Hz sine wave fields at 100 μT (electric field component not given) and detected changes in the transcription of key cellular genes (c-*fos*, c-*jun*, c-*myc*, protein kinase C). These changes, detected both by Northern blot analysis and by nuclear run-off assays, were dependent on the cell concentration used (5×10^5 or 10^6 ml^{-1}). At peak activities a 1.5–2-fold increase in transcription of c-*fos*, a change in transcription of c-*jun* (increase or decrease depending on cell concentration used), a 2–4-fold increase in transcription of c-*myc* and an up to three-fold increase in transcription of protein kinase C were found. However, in additional experiments a decrease rather than an increase in transcription of c-*fos* was noted, possibly suggesting that differences, perhaps quite subtle ones, in the actual state of the cells might be critical in influencing not only the scale of the effect, but also its direction. The time dependence of the effects was different for different mRNAs. Transcription of other genes (encoding metallothionein, transferrin, insulin receptor and ornithine decarboxylase, not normally expressed in these cells) was not affected. These workers '. . . interpret these data cautiously as indicating a more definite, non-random effect of magnetic field exposure on gene transcription'.

In an attempt to elucidate further the mechanism of the biological activity of EMF, Phillips *et al.* (1992b) evaluated the EMF-induced alterations in binding activities of several transcription factors using a gel-retardation assay. After 30 min exposure to a 100 μT, 60 Hz field (electric field component not specified) they found a 50–70% decrease in AP-1 binding activity accompanied by a decrease both in the mRNA for c-*jun* and in the c-*jun* protein. In further experiments (Phillips *et al.*, 1992a), they employed a series of *fos* promoter/CAT constructs in which various regulatory elements were singly deleted, and transfected them into CCRF-CEM cells. The aim was to determine which regulatory element is required for EMF-induced changes in c-*fos* transcription. They were particularly interested in the Ca^{2+}- and the cAMP-dependent regulatory elements, considering previous data in the literature that seemed to show an effect of EMF on both intracellular Ca^{2+} (Conti *et al.*, 1985; Carson *et al.*, 1990; Walleczek & Liburdy, 1990) and cAMP (Luben *et al.*, 1982; Farndale & Murray, 1986). However, their preliminary results did not indicate any role for these regulatory elements in EMF-induced changes in c-*fos* transcription.

Czerska *et al.* (1991) used the human promyelogenous leukaemia cell line HL-60 (the same cell line used by Goodman and co-workers) and a human lymphoma cell line (Daudi) to investigate the effects of 60 Hz magnetic fields at 75 μT (electric field component not specified) on transcript levels for c-*myc*. Cells were exposed for 30, 60 and 180 min. Using Northern blot analysis, they found a significant increase of c-*myc* mRNA in Daudi cells but no effect in HL-60. Similar results (Ning *et al.*, 1991) were obtained when they measured the total RNA content (by staining with acridine orange) in HL-60 and in Daudi cells that were exposed to a 60 Hz magnetic field of 100 μT (electric field component not specified) for 30 and 60 min.

Krause *et al.* (1992) measured levels of c-*myc* mRNA in both HL-60 and Daudi cells. They confirmed the results presented earlier by Czerska and co-workers, finding a 30% increase in total c-*myc* message after 90 min exposure of Daudi cells to a 60 Hz magnetic field of 100 μT (electric field component not specified). In complementary experiments, they employed an S-1 nuclease protection assay to determine relative rates of synthesis for c-*myc* mRNA. Again, they found an approximately 30% increase following 90 min exposure of Daudi cells to 60 Hz, 100 μT fields. When cells were exposed to 60 Hz fields of 1 mT, rates of transcription for c-*myc* were increased by 64%. Like Czerska and co-workers, they did not find any change when using HL-60 cells. However, these negative results with HL-60 cells contrast with those presented earlier by the same group (Krause *et al.*, 1989). In the earlier studies, they worked with 60 Hz magnetic fields of 1 mT (electric field: 1.5 mV m^{-1}) and reported a 2–3-fold increase in mRNA levels for c-*myc* and histone H2B, when HL-60 cells were exposed for 45–60 min to these fields.

In a comprehensive paper from the same group (Greene *et al.*, 1991) experiments using similar exposure parameters but measuring [^{3}H]uridine incorporation into RNA of HL-60 cells to assess the effects of EMF were described. A transient increase in [³H]uridine incorporation was found, which reached a maximum of 50–60% enhancement at 30–120 min exposure and declined to near basal levels by 18 h. The use of culture dishes with two concentric compartments enabled exposure of the cells to two different electric fields induced by the same magnetic field. For the inner compartment they calculated an electric field of 0.34 mV m⁻¹ whereas for the outer compartment it was 3.4 mV m⁻¹. The cells in the outer compartment exhibited greater incorporation of [³H]uridine (maximum of 50–70% increase) relative to controls than did cells in the inner compartment (maximum of 20–25% increase). Additionally, when cells were exposed to different magnetic fields (100 μT and 1 mT) inducing the same electric fields (0.34 mV m⁻¹), the enhancement in [³H]uridine incorporation was nearly identical (about 25%). These results indicate that the observed biological effects of magnetic fields could, at least in part, be due to the induced electric field.

The biological and field exposure aspects of the experiments of Greene and co-workers were duplicated by Azadniv & Miller (1992), in an attempt to reproduce the results. However, they were unable to detect any of the effects on [³H]uridine uptake which had been described.

Rehnolm *et al.* (1991) investigated normal (human amniotic fluid cells and peripheral blood lymphocytes) as well as malignant (embryonal carcinoma F9 and promyelocytic leukaemia HL-60) cells that were exposed to a 50 Hz, 30 μT field (electric field component not given) for 1–72 h. The rate of protein synthesis was measured as incorporation of [³⁵S]methionine during the last 15–30 min of exposure. The overall protein synthesis pattern was evaluated by means of two-dimensional gel electrophoresis. Although they did not see any significant change in [³⁵S]methionine incorporation, they noted a change in overall protein pattern after exposure to EMF both in normal and in malignant cells. These results are in agreement with the earlier work of Goodman & Henderson (1988).

More recent activities of the Goodman group, in cooperation with Blank and co-workers, have been focused on attempts to characterise further the biological response of cells to EMF and to elucidate whether or not there are common cellular reaction pathways in response to stress factors, similar to those used in response to EMF. For this purpose they have compared the response to EMF with those to certain stress factors such as heat shock (Blank, Khorkova & Goodman, 1992a). Salivary gland cells of *Sciara coprophila* were exposed to a heat shock (HS) of 37 °C (compare 25 °C normal growth condition). Alterations in HS-treated cells were then compared with those seen in cells exposed to 60 Hz magnetic fields of 0.57 to 570 μT (rms). The electric field component was not specified, but was probably of the order of 1 μV m^{-1} to 1 mV m^{-1} as stated elsewhere (Blank et al., 1992b). Alterations were evaluated by analysing patterns of 2D electrophoresis of [^{35}S]methionine-labelled, newly synthesised proteins. The authors reported that changes in protein patterns of HS-treated cells were very similar to those seen in cells exposed to EMF. They concluded that HS and EMF might affect a common final transduction pathway in cellular responses. They suggested that, since EMF-specific response pathways in cells may be deemed unlikely, EMF might be 'seen' by cells simply as an external stress factor. On the other hand, these results seem to contradict earlier observations reported by the same group (Goodman & Henderson, 1988; Goodman, 1991). In those studies, samples of salivary gland cells of Sciara coprophila were exposed to HS (37.5 °C) and, in parallel, to 60 Hz, 0.57 to 570 μT (rms) magnetic fields. Although some similarities in protein patterns between cells that were exposed to HS and those exposed to EMF were identified, the authors came to the conclusion that the overall protein pattern, the number of polypeptides resolved and the degree of augmentation were distinctly different between HS and EMF-exposed cells. Differences in cellular responses to HS and EMF are further supported by their other work on yeast (Goodman et al., 1992a) and HL-60 (Goodman, 1991). It would appear that further studies are necessary to resolve this apparent contradiction.

In order to test the idea that electric fields induced by applied magnetic fields might be responsible for the alterations seen in cells exposed to EMF, Goodman and Blank have completed a series of experiments where they have exposed HL-60 cells to an electric field applied directly via electrodes (Blank et al., 1992b). This same exposure system has been used by Blank to study the effects of weak sinusoidal alternating current on enzyme systems (Blank & Soo, 1989, 1990). Similar to the effects observed with EMF, Goodman and Blank noted an increase of transcript levels for histone H2B and c-myc in the range of 30 and 60%, respectively, above control levels and located windows for the effects in frequency, intensity and time. Maximal effects were seen at a frequency of 45 Hz (in the range used of 15–150 Hz) and at a field intensity of 3 mV m^{-1} (of the 0.3, 3, 30 and 300 mV m^{-1} that were investigated), while at the optimum field intensity, the current destiny through the cell suspension was 0.68 μA cm^{-2}. However, although these results are quite comparable to those reported with weak EMF, the optimum electric field at around 3 mV m^{-1} is two to three orders of magnitude higher than the induced electric fields calculated for the earlier experiments of Goodman and co-workers with EMF. In those experiments, the maximum stimulation was reported at a magnetic field of 5.7 μT (rms) for which the induced electric field was calculated to be $10 \, \mu V \, m^{-1}$ (Blank *et al.*, 1992b). Blank and Goodman did not include these much lower electric fields in their recent investigations, making it difficult to compare the results obtained with the two exposure systems. However, since the authors claim the existence of certain windows for the biological activity of both electric and electromagnetic fields, it could well be that there are different windows, one around $3 \, mV \, m^{-1}$ (in the absence of a magnetic field) and another one around $10 \, \mu V \, m^{-1}$ (in the presence of an appropriate magnetic field).

The reported biological effects of electric fields in the range of several millivolts per metre agree with earlier studies of Blank, who found an inhibition of Na/K-ATPase activity (Blank & Soo, 1989, 1990), as well as with findings of others such as Greene *et al.* (1989, 1991) and of Krause *et al.* (1989) as described above. Furthermore, a series of medical devices used for the treatment of bone fractures in humans imposes 2 mT pulsed magnetic fields, which induce current densities of about $1 \, \mu A \, cm^{-2}$ and associated electric fields of 1–$10 \, mV \, m^{-1}$in extracellular fluids (Luben *et al.*, 1982), but the therapeutic effectiveness of such devices is as controversial as the question of biological effects of low-intensity EMF.

On considering the results published so far on EMF-induced changes in gene expression, several major problems become evident. One of these is the apparent lack of coordinated studies involving the participation of different independent laboratories working under similar conditions with respect both to the biological system and to the electromagnetic environment. It would appear now that such coordinated approaches are going to be essential to answer convincingly the question of the possibility of biological effects of weak EMF at the cellular level. Another quite substantial problem is connected with the scale of the effects that have been reported so far. Most of them have been in the range of 20–70%, and only a few reports have shown changes at peak levels of more than twice the control values. The commonly used analytical methods for the quantification of gene expression (e.g. Northern blots, dot-blots, nuclear run-off assays) are possibly either not sensitive or not specific enough convincingly to demonstrate significant differences of the order of 20–50% between two cell samples. More effort evidently needs to be spent on improving these experimental methods or developing new more sensitive and reproducible assays.

One of the main reasons for the controversy about biological effects of weak EMF in general, is the lack of consensus on a definite mechanism that might explain these phenomena. The consequence is that the reported effects on cells or organisms exposed to EMF are viewed with caution and scepticism in the light of what is currently known about the

140   B. REIPERT

physical and biological state of cells. Adair, one who doubts the existence
of biological effects of weak EMF, states this unequivocally: '. . . there
are very good reasons to believe that weak electromagnetic fields can
have no significant biological effect at the cell level – and no strong
reason to believe otherwise' (Adair, 1991). His main argument is that,
if interactions of EMF at the cell level are to result in any significant
biological effects, those interactions must be significantly greater than
the ordinary thermal interactions of molecules with their environment.
He then calculates that magnetic fields in the range of tens of microteslas
would be too small to overcome the fundamental thermal noise in cell
systems. However, theoretical approaches by others seem to indicate that
it might be possible to conceive modes of interaction between EMF and
cellular systems to which the thermal noise limit might not necessarily
be an obstacle (Weaver & Astumian, 1990; Edmonds, 1993). On the
other hand, theoretical considerations, even when representing extra-
polations and interpolations of tested observations, cannot supplant a
well-established observation. In conclusion, if any biological effect of
weak EMF were to be established unambiguously, one would have to be
very concerned that these fields might in fact generate changes at the
cellular level which could have, if not direct carcinogenic potential, at
least cancer-promoting consequences.

References

Adair, R. K. (1991). Constraints on biological effects of weak
extremely-low-frequency electromagnetic fields. *Physical Review A*
43, 1039–48.
Azadniv, M. & Miller, M. W. (1992). ^{3}H-uridine uptake in human leukemia HL-60 cells exposed to extremely low frequency electromagnetic fields. *Biochemical and Biophysical Research Communications*
189, 437–44.
Blank, M., Khorkova, O. & Goodman, R. (1992a). Similarities in the
distribution of proteins following electromagnetic stimulation and
heat shock of *Sciara* salivary glands. *The Annual Review of Research
on Biological Effects of Electric and Magnetic Fields from the Generation, Delivery and Use of Electricity*. Abstract A-2, 8–11 November,
San Diego, California, USA.
Blank, M. & Soo, L. (1989). The effects of alternating currents on
Na,K-ATPase function. *Bioelectrochemistry and Bioenergetics* **22**,
313–22.
Blank, M. & Soo, L. (1990). Ion activation of the Na,K-ATPase in
alternating currents. *Bioelectrochemistry and Bioenergetics* **24**, 51–61.
Blank, M., Soo, L., Lin, H., Henderson, A. S. & Goodman, R.
(1992b). Changes in transcription in HL-60 cells following exposure

to alternating currents from electric fields. *Bioelectrochemistry and Bioenergetics* **28**, 301–9.

Carson, J. J., Prato, F. S., Drost, D. J., Diesbourg, L. D. & Dixon, S. J. (1990). Time-varying magnetic fields increase cytosolic free Ca^{2+} in HL-60 cells. *American Journal of Physiology* **259**, C687–92.

Coleman, M. P., Bell, C. M. J., Taylor, H. L. & Primic-Zakelj, M. (1989). Leukemia and residence near electricity transmission equipment: a case-control study. *British Journal of Cancer* **60**, 793–8.

Conti, P., Gigante, G. E., Alesse, E., Cifone, M. G., Fiesche, C., Reale, M. & Angeletti, P. U. (1985). A role for calcium in the effect of very low frequency electromagnetic fields on the blastogenesis of human lymphocytes. *FEBS Letters* **181**, 28–32.

Czerska, E., Al-Barazi, H., Casamento, J., Davis, C., Elson, E., Ning, J. & Swicord, M. (1991). Comparison of the effect of ELF field on c-*myc* oncogene expression in normal and transformed human cells. *13th Annual Meeting of the Bioelectromagnetic Society of the USA.* Abstract B-2-14, 23–27 June, Salt Lake City, Utah, USA.

Edmonds, D. T. (1993). Larmor precession as a mechanism for the detection of static and alternating magnetic fields. *Bioelectrochemistry and Bioenergetics* **30**, 3–12.

Farndale, R. W. & Murray, J. C. (1986). The action of pulsed magnetic fields on cyclic AMP levels in cultured fibroblasts. *Biochimica et Biophysica Acta* **881**, 46–53.

Feychting, M. & Ahlbom, A. (1992). Magnetic fields and cancer in people residing near Swedish high voltage power lines. *IMM-Rapport* **6/92**, Karolinska Institute, Stockholm, Sweden.

Goodman, R. (1991). Transcription and translation in cells exposed to extremely low frequency electromagnetic fields. *Bioelectrochemistry and Bioenergetics* **25**, 335–55.

Goodman, R., Abbott, J. & Henderson, A. S. (1987). Transcriptional patterns in the X chromosomes of *Sciara coprophila* following exposure to magnetic fields. *Bioelectromagnetics* **8**, 1–7.

Goodman, R., Bassett, C. A. L. & Henderson, A. S. (1983). Pulsing electromagnetic fields induce cellular transcription. *Science* **220**, 1283–5.

Goodman, R. & Henderson, A. S. (1988). Exposure of salivary gland cells to low-frequency electromagnetic fields alters polypeptide synthesis. *Proceedings of the National Academy of Sciences USA* **85**, 3928–32.

Goodman, R., Khorkova, O., Henderson, A. & Weisbrot, D. (1992a). The effect of low frequency electric and magnetic fields on gene expression in *Saccharomyces cerevisiae*. *The Annual Review of Research on Biological Effects of Electric and Magnetic Fields from the Generation, Delivery and Use of Electricity*. Abstract P-15, 8–11 November, San Diego, California, USA.

Goodman, R., Wei, L. X. & Henderson, A. (1991). An increase in mRNA and protein levels is induced in human cells exposed to electromagnetic fields: relationship to time of exposure, post exposure time and frequency. *The Annual Review of Research on Biological Effects of 50 and 60 Hz Electric and Magnetic Fields.* Abstract A-18, 3–7 November, Milwaukee, Wisconsin, USA.

Goodman, R., Wei, L.-X., Xu, J.-C. & Henderson, A. (1989). Exposure of human cells to low-frequency electromagnetic fields results in quantitative changes in transcripts. *Biochimica et Biophysica Acta* **1009**, 216–20.

Goodman, R., Weisbrot, D., Uluc, A. & Henderson, A. (1992b). Transcription in *Drosophila melanogaster* salivary gland cells is altered following exposure to low-frequency electromagnetic fields: analysis of chromosome 3R. *Bioelectromagnetics* **13**, 111–18.

Greene, J. J., Skrowronski, W. J., Krause, D., Mullins, J. M. & Nardone, R. M. (1989). Delineation of the magnetic and electric field contributions to the enhancement of transcription by low frequency radiation. *11th Annual Meeting of the Bioelectromagnetic Society of the USA.* Abstract P-1-12, 18–22 June, Tucson, Arizona, USA.

Greene, J. J., Skrowronski, W. J., Mullins, J. M. & Nardone, R. (1991). Delineation of electric and magnetic field effects of extremely low frequency electromagnetic radiation on transcription. *Biochemical and Biophysical Research Communications* **174**, 742–9.

Krause, D., Brent, J. A., Mullins, J. M., Greene, J. J. & Nardone, R. M. (1989). Induction of c-*myc* and histone H2B in HL-60 cells by extremely low frequency (ELF) electromagnetic radiation (EMR). *11th Annual Meeting of the Bioelectromagnetic Society of the USA.* Abstract P-1-11, 18–22 June, Tucson, Arizona, USA.

Krause, D., Greene, J., Desta, A., Pearson, S., Nardone, R. M. & Mullins, J. M. (1992). 60 Hz magnetic field exposure alters the response of cultured cells for ornithine decarboxylase activity and c-*myc* mRNA. *The First World Congress for Electricity and Magnetism in Biology and Medicine.* Abstract P-141, 10–14 June, Lake Buena Vista, Florida, USA.

London, S. J., Thomas, D. C., Bowman, J. D., Sobel, E., Cheng, T. C. & Peters, J. M. (1991). Exposure to residential electric and magnetic fields and risk of childhood leukaemia. *American Journal of Epidemiology* **134**, 923–38.

Luben, R. A., Cain, C. D., Chen, M. C.-Y., Rosen, D. M. & Adey, W. R. (1982). Effects of electromagnetic stimuli on bone and bone cells *in vitro*: inhibition of responses to parathyroid hormone by low-energy low-frequency fields. *Proceedings of the National Academy of Sciences USA* **79**, 4180–4.

National Radiological Protection Board (NRPB) (1992a). Electromagnetic fields and the risk of cancer. Report of an advisory group on non-ionising radiation. *Documents of the National Radiological Protection Board* **3**, 19–20.

National Radiological Protection Board (NRPB) (1992b). Electromagnetic fields and the risk of cancer. Report of an advisory group on non-ionising radiation. *Documents of the National Radiological Protection Board* 3, 23.

Ning, J., Al-Barazi, H., Casamento, J., Czerska, E., Davis, C., Elson, E. & Swicord, M. (1991). Comparison of the effect of ELF fields on total RNA content in normal and transformed human cells. *13th Annual Meeting of the Bioelectromagnetic Society of the USA*. Abstract B-2-13, 23–27 June, Salt Lake City, Utah, USA.

Phillips, J. L., Haggren, W., Thomas, W. J., Ishida-Jones, T. & Adey, W. R. (1992a). Effects of 60 Hz magnetic fields exposure on c-fos transcription in CCRF-CEM human T-lymphoblastoid cells. *The First World Congress for Electricity and Magnetism in Biology and Medicine*. Abstract O-5, 10–14 June, Lake Buena Vista, Florida, USA.

Phillips, J. L., Haggren, W., Thomas, W. J., Ishida-Jones, T. & Adey, W. R. (1992b). Effects of 60 Hz magnetic field exposure on transcription factor AP-1 binding activity in human T-lymphoblastoid cells. *The Annual Review of Research on Biological Effects of Electric and Magnetic Fields from the Generation, Delivery and Use of Electricity*. Abstract A-3, 8–11 November, San Diego, California, USA.

Phillips, J. L., Haggren, W., Thomas, W. J., Ishida-Jones, T. & Adey, W. R. (1992c). Magnetic field-induced changes in specific gene transcription. *Biochimica et Biophysica Acta* 1132, 140–4.

Phillips, J. L. & McChesney, L. (1991). Effect of 72 Hz pulsed magnetic field exposure on macromolecular synthesis in CCRF-CEM cells. *Cancer Biochemistry and Biophysics* 12 1–7.

Rehnholm, U., Johnsson, C., Mild, K. H. & Mattsson, M.-O. (1991). Altered pattern of protein synthesis in cultured cells after magnetic field exposure. *13th Annual Meeting of the Bioelectromagnetic Society of the USA*. Abstract B-2-8, 23–27 June, Salt Lake City, Utah, USA.

Savitz, D. A., Wachtel, H., Barnes, F. A., John, E. M. & Tvrdick, J. G. (1988). Case-control study of childhood cancer and exposure to 60 Hz magnetic fields. *American Journal of Epidemiology* 128, 21–38.

Walleczek, J. & Liburdy, R. P. (1990). Nonthermal 60 Hz sinusoidal magnetic field exposure enhances $^{45}Ca^{2+}$ uptake in rat thymocytes: dependence on mitogen activation. *FEBS Letters* 271, 157–60.

Weaver, J. C. & Astumian, R. D. (1990). The response of living cells to very weak electric fields: the thermal noise limit. *Science* 247, 459–62.

Wei, L-X., Goodman, R. & Henderson, A. (1990). Changes in level of c-*myc* and histone H2B following exposure of cells to low-frequency sinusoidal electromagnetic fields: evidence for a window effect. *Bioelectromagnetics* 11, 269–72.

Wertheimer, N. & Leeper, E. (1979). Electrical wiring configurations and childhood cancer. *American Journal of Epidemiology* 109, 273–86.

Part 2
Hard tissue

A. J. EL HAJ and G. P. THOMAS

Cellular modelling of mechanical interactions with the skeleton

Introduction

The role of exercise and mechanical stimuli in promoting soft tissue hypertrophy has been well studied, with many aspects covered in this volume; the role of exercise and mechanical loading in defining skeletal hypertrophy and remodelling has been less well characterised. In both tissues the mechanical events which occur during exercise can be analysed and modelled in a series of experimental manipulations *in vivo* and *in vitro*.

Lanyon, in a later chapter, outlines experiments to alter the mechanical environment *in vivo* and describes an elevation in bone modelling in response to a changing load environment. In parallel experiments, immobilisation and neurectomy is linked with disuse atrophy of the skeleton. This adaptation of bone can be defined as the formation and resorption of bone tissue in response to load in order to achieve the capacity for a specific mechanical function.

In vivo models allow loading responses to be monitored within the *in situ* three-dimensional bone matrix environment. However, these experiments can often pose difficulties in isolating specific cellular events occurring in complex tissues, relying heavily on technically demanding histological analysis of hard tissues. Such problems in treatment and isolation of cellular responses *in vivo* have hindered the progress in attempts to define the effects of loading on bone cell growth and differentiation. *In vitro* models offer different advantages; for example, effects on specific cells can be studied, a problem in the heterogeneous *in vivo* environment. Cell culture allows further potentially load responsive characteristics such as morphology, density dependence, motility, adhesiveness and mortality to be investigated. McKeehan *et al.* (1990) in his review on mammalian

Society for Experimental Biology Seminar Series 54: *Biomechanics and Cells*, ed. F. Lyall & A. J. El Haj. © Cambridge University Press 1994, pp. 147–163.

cell culture have proposed that *in vitro* models should aim to fulfil a series of criteria:

1   the isolation of specific cells or tissues that retain specific properties observed *in vivo*;
2   the characterisation of the environmental factors which directly effect cell or tissue behaviour;
3   the reconstruction of the cell and its environment into a defined, experimentally manageable unit enabling the study of cell responses and properties in a dynamic context.

In addition with bone cells, a number of specific points can be made (El Haj, 1990):

4   to enable the cells to construct a 3-D matrix of collagenous and non-collagenous proteins similar to bone *in vivo*;
5   to identify and characterise the strain profile which is applied to an *in vitro* culture in order to mimic closely those experience *in vivo*.

In this chapter, key components of *in vivo* strains which induce bone tissue remodelling and modelling will be outlined (further detail available in following chapters) and a synopsis of various *in vitro* strain models. Finally, results from two recent models, a compressive macroporous bead model and a marrow stromal cell stretched membrane model, will be discussed.

What is strain?

The field of biomechanics can often appear complex and analysis reliant upon difficult mathematical formulae. These complex distributions of strain are often difficult to model accurately *in vitro*. The application of mechanical force (or load) can be simply described as a deformation of shape or dimensions of a material (strain). The force within the material resisting this deformation is then termed stress (Frost, 1988). Strain has been defined as the change in length per unit length and is thus dimensionless. As a result, arbitrary units, Strains (str), are used to define length changes; one Strain is a change in length of 100%. In physiological systems, units of microstrains (μstr) are more commonly used, 1 μstr representing a change in length of 0.0001%.

Three forms of strain have been defined: tensile, compressive and shear. All three are experienced by bone cells during the everyday activities of an organism. Strains most commonly experienced in bone matrix are compressive and tensile strains. Shear makes up a varying component of the total strain. Theoretically bones would adapt so these strains would be applied axially, therefore minimising the strains so all cortices experience longitudinal compression. However, this has been shown to be the case for only one bone studied, the metacarpus (Biewener, Thomason & Lanyon, 1983). Other load-bearing bones show a curved morphology with the strain axis passing through the marrow cavity. This places very low strains on the cortices through which the axis passes (the epiphysis). Longitudinal compression and longitudinal tension are experienced by the other two cortices, i.e. bending. Bending has been calculated to be responsible for up to 89% of total strain in bone (Rubin, 1984).

In vivo strain magnitudes

Lanyon (1984) has shown that the maximal strain normally experienced by load-bearing bones is 1500–3000 μstr. Rubin (1984) has published a table of maximal strains experienced by parts of the skeleton in various animals measured using strain gauge analysis. Fracture strain in longitudinal tension of compression has been shown to be in the region of 25 000 μstr. This allows a 10-fold safety factor before fracture occurs. Long-term fatigue studies have shown that bone subjected to average strains of 2000 μstr has a fatigue life in excess of 10^6 cycles. Increasing the average strain to above normal levels (4000 μstr) reduces the fatigue life-span to approximately 20 000 cycles. Considering that the lower extremities experience 1–10 million cycles a year this is a very short life-span. Frost (1988) has proposed a mechanism whereby a minimum effective strain (MES) controls the modelling and remodelling responses of the bone. Strain levels of 100–300 μstr result in a remodelling response maintaining bone mass. Modelling is triggered by strains of 1500–3000 μstr and results in the depression of remodelling and increases in cortical bone. Data showing a proportional increase in bone mass with graded dynamic loads above 1000 μstr support this suggestion (Rubin & Lanyon, 1984). Lanyon (1987) proposes a minimum effective strain-related stimulus (MESS) which takes into account strain magnitude, rate and distribution as well as non-mechanical influences such as stress and hormones. Strain distribution, especially, may influence the loading response. Thus, if a bone is adapted to a particular strain and the axis of the strain alters the bone will need to adapt to the new stresses imposed upon it even though the magnitude may be constant (Lanyon, 1987). Lanyon elaborates these theories in his chapter, including work on cellular detection of strain *in vivo*.

In vivo, most strains commonly experienced are of a cyclic nature, such as those imposed by walking or flying. Experimental studies therefore have focused on the application of cyclic strains. Intermittent compressive strains rather than static loads have been shown to induce the greatest response. Hypertrophy induced by cyclic loading applied by normal locomotion was greater than that induced by static compressive loads on immobilised limbs in dogs (Chamay & Tschantz, 1972). Cyclic loading rates of 0.5–1 Hz have been used because of their proximity to that of walking, i.e. normal physiological rates.

Duration of load has been shown to have a minimal effect on the response in bone. As few as 4 cycles per day (0.5 Hz, 8 s) were shown to arrest negative remodelling and a maximal response was seen after only 36 cycles (Rubin & Lanyon, 1984). Increased bone formation has also been seen in response to a single period of loading lasting only 300 cycles (1 Hz) (Pead, Skerry & Lanyon, 1988).

The difficulty arises when trying to interpret these *in vivo* studies into an effective *in vitro* model. The lack of accurate measurements of the strains which a cell may be exposed to *in vivo* poses problems with establishing a physiological model of cellular strains. Criticism is often levelled at the fact that without a matrix surrounding the cells *in vitro*, models do not present an effective comparison with events occurring *in situ* and that the complex strains are often oversimplified. Cowin, Moss-Salentijn & Moss (1991) have suggested that lacunar shape and localised structural anisotropy may increase tissue level strains by 10-fold at the cellular level. Others have suggested that the cells may be experiencing much lower levels of strain than the surrounding matrix. Nevertheless, cell models have been used extensively to attempt to address many questions related to mechanically induced bone modelling and their findings have gone some way to identify possible mechanotransduction mechanisms for strain and to identify load responses.

In vitro strain magnitudes

When applying loads to cells in culture the differences between the type of strain, compressive versus tensile is not straightforward. Using a balloon as a model for a cell, compressive and tensile forces have been shown to have the same end result on the volume (Jones & Bingmann, 1991). Tensile or compressive force results in length increases in one plane but decreases in the opposite plane maintaining the volume. Whether these different forces have dose–response transduction effects in terms of growth factor production or secondary messenger systems has yet to be determined. Cell culture models involving stretching, bending or expanding the cell substrate are the most common and easily constructed method of applying tensile/compressive force. Jones elaborates two of his models, the stretch sheet model and the four-point bending model, in his chapter. A new model involving compression of beads seeded with cells has recently been developed and will be covered later in this chapter. These experiments simulate the compressive strains experienced by bone through the use of an artificial matrix composed of macroporous gelatin beads.

An alternative form of compressive force is hydrostatic pressure, used by Burger and colleagues who outline their system further in the following chapter. Again though minimal changes in volume will be seen due to the incompressibility of liquids (cell contents), shear strains are created when two materials move past each other.

The magnitude of strain applied to bone cells *in vitro* covers a very broad range. Many studies have made no effort to calibrate the strain experienced by the cells and only recorded the force applied. However, the strain experienced by cells to a particular force depends upon the nature of the loading system and comparisons between systems can often be difficult or misleading. In many studies both non-physiological forces are being applied and strains experienced.

Studies applying hydrostatic compressive force, such as those of Klein Nulend and associates, have only calibrated the force applied. Pressures of 13 kPas were sufficient to induce osteogenic responses (Burger, Klein Nulend & Veldhuijzen, 1992). The compressive force applied by Lozupone, Favia & Grimaldi (1992) using a metatarsal organ culture model was calculated to be twice that experienced when standing, resulting in increased osteoid thickness and osteoblast proliferation as well as the number of viable osteocytes. Duncan & Misler (1989) described the activation of a stretch-activated channel opened by a suction of > 5 mm Hg by applying a suction via patch clamp. Using a micromanipulator, McDonald & Houston (1992) calculated that they were applying an average strain of 65 μstr when applying suction to a cell. Calculations of strain at such a microscopic level require possibly inaccurate estimates of several factors and therefore must be viewed cautiously (Cowin *et al.*, 1991; Sachs, 1988). A pressure of 10 kg cm^{-2} induced a deformation in length of 0.1–0.05% (equivalent to 1000–5000 μstr) in the system used by Binderman *et al.* (1988), which is reviewed in a later chapter.

Further investigations have used fluid flow to induce shear force, a situation thought to occur *in vivo* (Reich & Frangos, 1991). Frangos, Eskin & McIntire (1988) calculated that shear stresses from 1 to 24 dyn cm^{-2} were experienced physiologically and subsequently subjected calvarial osteoblast cells to stresses at these levels, resulting in elevated secondary messenger activity. Apart from the very low strains claimed to be applied in patch-clamping, the lowest documented strains to exert an effect on bone cells are between 400 and 700 μstr. A longitudinal compressive force of 650 μstr at 0.4 Hz was applied by Zaman, Dallas & Lanyon (1992) on embryonic chick tibia in culture, resulting in elevated glucose 6-phosphate dehydrogenase (G6PD) activity, collagen and alkaline phosphatase synthesis. Brighton *et al.* (1991) saw increases in DNA content and prostaglandin E_2 (PGE_2) production in response to tensile strains of 400 μstr applied at 1 Hz in rat calvarial osteoblasts. No response was seen when strains of 200 or 1000 μstr were applied, however. Jones *et al.* (1991) demonstrated that bovine periosteal osteoblasts showed no response to cyclic tensile and compressive strains of 300 μstr. Cell division, collagen synthesis, collagenase activity and prostaglandin synthesis all increased when strains of 3000 μstr were applied, however (Jones *et al.*, 1991). Several other studies have shown responses to strains between 1000 and 10000 μstr. In the canine trabecular core *in vitro* model, a bulk compressive strain of 5000 μstr across the core was applied (El Haj *et al.*, 1990) resulting in elevated RNA and enzyme activity in the endosteum. A later study by Brighton *et al.* (1992), using the system mentioned above, showed increases in DNA and secondary messenger production in response to a strain of 1700 μstr. Hasegawa *et al.* (1985), in their expandable membrane model, used tensile strains of 4000 μstr, and Yeh & Rodan (1984) in their expandable collagen strip used tensile strains of between 1000 and 5000 μstr. A biphasic response was demonstrated by Murray & Rushton (1990); PGE_2 was increased in response to tensile strains of 7000 and 28 000 μstr but only a minimal response was seen at strains between these levels. Increases in cell division and alignment of osteoblasts has been measured (Buckley *et al.*, 1988) as well as an increase in osteoblastic activity (Buckley, Banes & Jordan, 1990) in response to strains of up to 24 000 μstr.

A non-osteogenic response has also been seen in some studies. When high non-physiological strains, tensile strains of 10 000 μstr, are applied to bone cells, osteoblasts assume a fibroblastic phenotype (Jones *et al.*, 1991; Jones & Bingmann, 1991). Increased levels of collagen type III, characteristic of fibroblasts, were produced by osteoblasts strained at this level. Synthesis of large molecular weight proteoglycans has also been reported (Jones & Bingmann, 1991). Synthesis of these products may be involved in the formation of a fibrous capsule in response to the large shear stresses induced by non-physiological loading. Other researchers have also shown catabolic responses to strains supposedly approaching the yield strain of bone *in vivo*.

As well as the magnitude, the nature of the load applied varies considerably. Loads have been applied for different durations either continually or intermittently. The frequency of the intermittent loads applied also varies. Klein Nulend *et al.* (1987a,b) have stated that discontinuous loading exerts a greater response in calcifying cartilage than a continuous load. Hasegawa *et al.* (1985) in their stretched membrane model found no difference in responses of continuously or intermittently stretched cells. Many researchers have chosen 1 Hz as the cycle frequency because of its parity with strain frequency experienced during locomotion; however, cycle time of intermittent loads appears to have little effect (Murray & Rushton, 1990; Jones *et al.*, 1991). The duration of load also appears to have little effect on the response. The majority of studies load for periods up to 1 h. Jones *et al.* (1991) have shown that only one cycle a day could increase secondary messenger levels.

Derivation of bone cells

Finally, it is important to consider where in the body the bone cells are derived from when interpreting cell culture models. Whereas for hormone experiments, where the influences will be circulating in a systemic fashion, loading may be isolated to regions of the body with different parts of the skeleton responding in different ways. This point is discussed further in the chapter by Skerry *et al.* The calvarial derived periosteal/ endosteal cells are the most widely used culture system available, providing sources of pre-osteoblasts through to mature osteoblast populations which can be separated crudely into populations by sequential digestion (Wong & Cohn, 1974) or stripping of calvaria (Matsuda & Davies, 1987). However, questions are raised as to whether calvaria experience strain-induced modelling or remodelling in the lifetime of an animal. Also, most studies are carried out using cells derived from embryonic and neonatal animals.

An alternative bone cell culture system is the bone marrow stromal cell, which provides a good system for studying bone cell lineage and differentiation of the mature osteoblast in culture. It is possible to extract the bone marrow from a long bone of an adult rat which will be experiencing strain throughout the life of the animal. Although the best derivation of a bone cell culture would be sourced from the periostea of long bones, stripping periostea from long bones is difficult when the aim is to achieve pure populations of bone cells without contamination from muscle cells and other cell types. Bovine cells derived from long bones have been used by Jones & Bingmann (1991).

A number of studies have used bone cell lines which are derived from osteosarcomas of rat or mouse. However, these cell lines may have different secondary messenger systems as a feature of the immortalisation

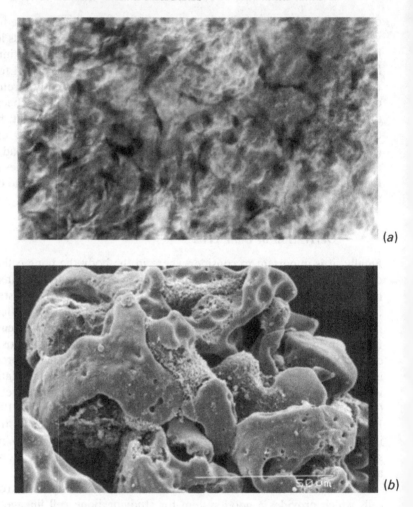

Fig. 1. (a) Photomicrograph of macroporous beads seeded with calvarial periosteal cells isolated by the method of Matsuda & Davies (1987). Stained with 0.1% toluidine blue in whole mount. Dark regions denote the cells within the gelatin matrix. (b) Scanning electron micrograph of a macroporous bead. Cells can be seen located within the pore structure.

process and have been shown to respond differently to treatment in culture (Shelton & El Haj, 1991).

In our studies, we have developed two models which attempt to address some of the many criticisms raised against cell cultures. In particular, the macroporous bead model establishes a synthetic gelatin matrix for the cells to grow in and form a trabecular network (Shelton & El Haj, 1991, 1992) (Fig. 1), and the bone marrow stromal cell model derived from long bone of adult rats to investigate effects of loading on bone cell differentiation.

Macroporous bead model

When left in culture for 3–4 weeks, bone cells will grow in macroporous gelatin beads (Fig. 1*a,b*) (Hyclone Ltd), eventually producing matrix proteins and collagen, and mineralising the beads which surround them (A. J. El Haj, unpublished data). Using this model, compressive load can be applied to a column of beads seeded with bone cells derived from differing sources and compressed in a cyclical fashion similar to the bone core model for trabecular bone developed by El Haj *et al.* (1989). The loads have been calibrated using a linear volume displacement transducer attached to a pneumatic piston and loads of up to 5000 μstr are applied across the column of beads analogous to the trabecular core experiments (El Haj *et al.*, 1990; Shelton & El Haj, 1992). Changes in hydrostatic pressure of the medium within the column are measured by cannulating the flow and using a pressure transducer. Compressive loading of the system has been demonstrated not to change the profile of measurements recorded during continuous perfusion without loading. In this way, isolated populations of bone cells from different parts of the skeleton can be grown on beads and compressively loaded in a matrix environment. The degree of matrix formation and mineralisation can be controlled by the time in culture.

Initially, calvarial derived periosteal cells have been used for the development of this model to form a comparison with previous work. Periosteal cells derived by the method of Matsuda & Davies (1987) are grown into the macroporous matrix. In response to compressive loading at 1 Hz, the cells demonstrate an elevation in the rate of RNA synthesis but not DNA synthesis (Fig. 2*a,b*) (Shelton & El Haj, 1992). This seems to indicate that cells are responding to compressive loading by elevating synthetic rates but not cellular proliferation. Elevated levels of mRNA have been measured in compressed bead columns for collagen type 1 and alkaline phosphatase (R. M. Shelton and A. J. El Haj, personal communication). Loading in the presence of indomethacin (10^{-6} M) blocks these responses (Shelton & El Haj, 1992).

Bone marrow stromal cell model

In parallel with this work, we have demonstrated that bone marrow stromal cells derived from rat long bones are load-responsive. Initially

156   A. J. EL HAJ & G. P. THOMAS

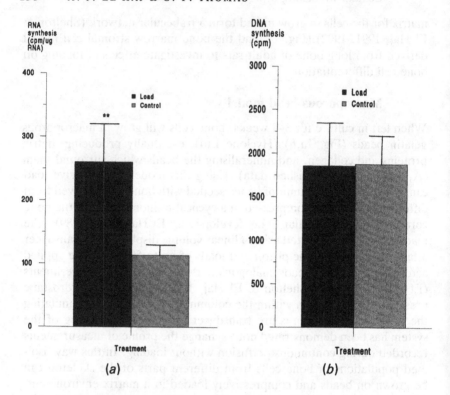

Fig. 2. Effect of compressive loading on periosteal cells grown on macroporous beads. (a) RNA synthesis in response to loading: 1 μCi ml⁻¹
[³H]uridine uptake over 16 h following a ½ h compressive loading at
1 Hz. Total RNA is extracted using the guanidium isothyanate/phenol
method, quantified spectrophotometrically and assayed by scintillation
counting ($n = 9$). (b) DNA synthesis in response to loading: [³H]thymid-
ine uptake following 16 h post loading. DNA was extracted using 2%
perchloric acid and solubilised in NAOH before scintillation counting
($b = 6$).

employing a tensile model using expandable Petriperm membrane dishes,
we have adapted a bone marrow culture model developed by Manioto-
polous, Sodek & Melcher (1988) to investigate the effects of load on
bone cell differentiation from stromal cells. Selection of stromal cells in
bone marrow cultures requires the removal of the non-adherent haemo-
poietic cells early on in culture. Media are changed after 24 h and subsequently every 48 h. With time in culture, bone marrow cells on Petriperm substrates will multilayer forming nodules which eventually

mineralise. This culture system then meets the criteria of cells derived from long bones and is capable of producing a matrix *in vitro*. Using this culture system, we can investigate differentiation of the osteoblastic phenotype from the stromal cell population and identify the effects of strain on this differentiation and cellular proliferation of precursors.

The effect of dexamethasone (dex), a known osteoblastic inducing agent, has been shown to increase bone cell differentiation and/or activity using FACS analysis. In response to dex (10^{-8} M), there is an elevation in the number of cells expressing alkaline phosphatase, osteopontin, and collagen type 1 and the amount produced in culture (Publicover, Thomas & El Haj, 1994). Changes in the type of Ca^{2+} membrane channel expressed have also been shown during differentiation of these bone cells in response to dexamethasone (Publicover *et al.*, 1994).

Our studies have been investigating whether it is possible to induce similar osteoblastic differentiation in response to loading. A single tensile pulse of 250 µstr at 1 Hz for ½ h at day 10 in culture results in elevations in alkaline phosphatase using FACS analysis 2 days later (Fig. 3). These results show an elevation in production of alkaline phosphatase and an increase in the number of cells expressing alkaline phosphatase. Levels of alkaline phosphatase expression are responsive to the levels of loading applied, with the greatest response occurring at 200–300 µstr. Loading at 10 days also results in an elevation in collagen mRNA production 2 days later (Fig. 4); however, increased protein levels are not measurable at this time. Our ongoing studies would seem to provide evidence that intermittent loading of low magnitude may promote increases in the number of cells which differentiate into an osteoblastic phenotype in isolated bone marrow cultures.

Bone marrow stromal cells can also be grown on macroporous beads with and without the presence of dex, thus providing two sources of cell populations, stem cells and dex-induced bone cells. These populations can then be loaded and the effects of compression in the presence of a gelatin matrix assessed. This work has demonstrated that dex-induced mature cell cultures elevate their RNA synthesis rates by two-fold, similar to that found in mature periosteal derived cells, but the nature of this elevation has not yet been characterised. Investigations into the effects of compressive loading on differentiation of stem cells are being carried out.

Our results have provided evidence to support the theory that compressive and tensile loads *in vitro* seem to promote similar responses in bone cells. However, we suggest that there are differences in the way bone cells at different stages of differentiation respond to mechanical loading (Fig. 5). Our studies have demonstrated that bone cells at all

158   A. J. EL HAJ & G. P. THOMAS

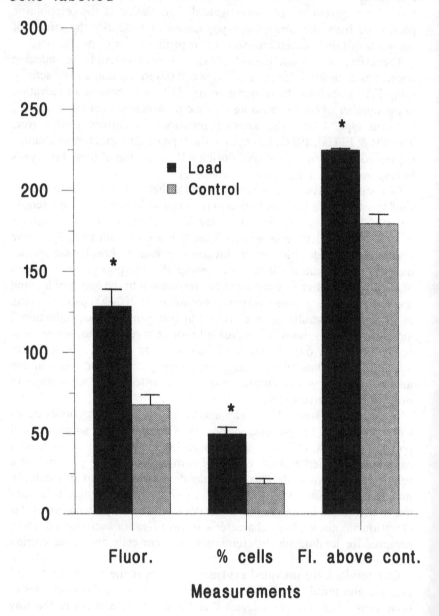

Fluor. or %
cells labelled

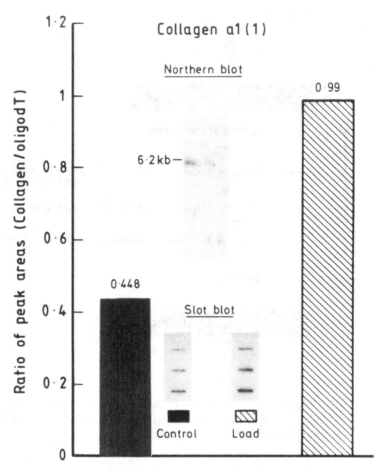

Fig. 4. Effect of tensile loading on collagen mRNA levels in bone marrow cultures. RNA extraction was carried out as described above followed by Northern gel analysis using 10 µg RNA per well, blotting on nitrocellulose gel and hybridisation with a cDNA probe for human collagen α1(1). Slot blot analysis was expressed in terms of oligodT hybridisation for total RNA and scanned densitometrically.

Fig. 3. The effects of tensile loading on the levels of alkaline phosphatase expression in bone marrow stromal cells: FACS analysis was used following treatment with a primary antibody for alkaline phosphatase (donated by G. Rodan). Fluor.: overall mean fluorescence per cell of 10 000 cells counted; % cells: percentage of cells binding alkaline phosphatase primary antibody out of 10 000 analysed; Fl. above cont.: mean level of fluorescence above control. Control level was set using a non-binding antibody and the absence of a primary antibody (n = 6 loaded and 6 controls).

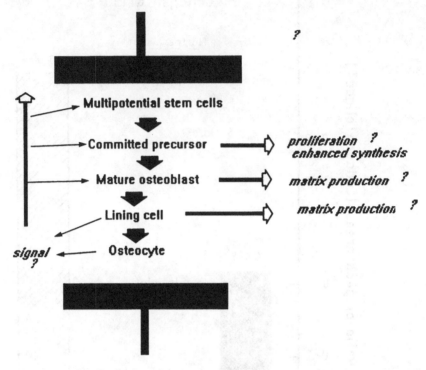

Fig. 5. Schematic diagram of the load effects on different stages of bone cell differentiation. Arrows denote the possible routes for each stage following intermittent mechanical loading.

stages of differentiation are load responsive. However, from our initial findings, we can hypothesise that loading affects more mature bone cells by promoting an elevation in matrix synthesis and cellular activity, whereas loading effects on bone cell stem and transit cells is to promote differentiation and not proliferation. There may be some proliferation of the committed precursor cell which would rely on other factors providing mitogenic signals for bone cell precursors and not mechanical stimuli. An interesting feature of most of the *in vitro* models discussed in this chapter is the absence of the osteocyte, the cell which has been put forward as a potential load-responsive cell by some investigators. It may be that osteocyte is producing various mitogenic signals which are inducing the proliferative response. However, even with data awaiting on the responses of the osteocyte *in vitro*, there is a mounting body of evidence that has shown that many stages of bone cell are load-responsive. Bone cells may be responding to a load stimulus in a dose-dependent manner according to the differentiation stage.

References

Biewener, A. A., Thomason, J. & Lanyon, L. E. (1983). Mechanics of locomotion and jumping in the forelimb of the horse. *Journal of Zoology (London)* **201**, 67–82.

Binderman, I., Zor, U., Kaye, A. M., Shimshoni, Z., Harell, A. & Somjen, D. (1988). The transduction of mechanical force into biochemical events in bone cells may involve the activation of phospholipase A2. *Calcified Tissue International* **42**, 261–6.

Brighton, C. T., Sennett, B. J., Farmer, J. C., Iannotti, J. P., Hansen, C. A., Williams, J. L. & Williamson, J. (1992). The inositol phosphatase pathway as a mediator in the proliferative response of rat calvarial bone cells to cyclical biaxial mechanical strain. *Journal of Orthopaedic Research* **10**, 385–93.

Brighton, C. T., Strafford, B., Gross, S. B., Leatherwood, D. F., Williams, J. L. & Pollack, S. R. (1991). The proliferative and synthetic response of isolated calvarial bone cells of rats to cyclical biaxial mechanical strain. *Journal of Bone and Joint Surgery* **73A**, 320–31.

Buckley, M. J., Banes, A. J. & Jordan, R. D. (1990). The effects of mechanical strain on osteoblasts *in vitro*. *Journal of Oral and Maxillofacial Surgery* **48**, 276–82.

Buckley, M. J., Banes, A. J., Levin, L. G., Sumpio, B. E., Sato, M., Jordan, R., Gilbert, J., Link, G. W. & Tran Son Tay, R. (1988). Osteoblasts increase their rate of division and align in response to cyclic mechanical tension *in vitro*. *Bone and Mineral* **4**, 225–36.

Burger, E. H., Klein-Nulend, J. & Veldhuijzen, J. P. (1992). Mechanical stress and osteogenesis *in vitro*. *Journal of Bone and Mineral Research* **7**(S2), S397–401.

Chamay, A. & Tshantz, P. (1972). Mechanical influences in bone remodelling. Experimental research on Wolff's Law. *Journal of Biomechanics* **5**, 173–80.

Cowin, S. C., Moss-Salentijn, L. & Moss, M. L. (1991). Candidates for mechanosensory system in bone. *Journal of Biomechanical Engineering* **113**, 191–7.

Duncan, R. L. & Misler, S. (1989). Voltage activated and stretch activated Ca^{2+} conducting channels in an osteoblast like cell line (UMR106). *FEBS Letters* **251**, 17–21.

El Haj, A. J. (1990). Biomechanical bone cell signalling: is there a grapevine? *Journal of Zoology (London)* **220**, 689–93.

El Haj, A. J., Minter, S. L., Rawlinson, S. C. F., Suswillo, R. & Lanyon, L. E. (1990). Cellular response to mechanical loading *in vitro*. *Journal of Bone and Mineral Research* **5**, 923–32.

Frangos, J. A., Eskin, S. G. & McIntire, L. V. (1988). Shear stress induced mammalian cell stimulation. *Biotechnology and Bioengineering* **32**, 1053–60.

Frost, H. M. (1988). Vital biomechanics: proposed general concepts for skeletal adaptation to mechanical usage. *Calcified Tissue International* **42**, 145–56.

Hasegawa, S., Sato, S., Saito, S., Suzuki, Y. & Brunette, D. M. (1985). Mechanical stretching increases the number of cultured bone cells synthesizing DNA and alters their pattern of protein synthesis. *Calcified Tissue International* **37**, 431–6.

Jones, D. B. & Bingmann, D. (1991). How do osteoblasts respond to mechanical stimulation? *Cells and Materials* **1**, 329–40.

Jones, D. B., Nolte, H., Scholubbers, J.-G., Turner, E. & Veltel, D. (1991). Biochemical signal transduction of mechanical strain in osteoblast-like cells. *Biomaterials* **12**, 101–9.

Klein-Nulend, J., Veldhuijzen, J. P., De Jong, M. & Burger, E. H. (1987a). Increased bone formation and decreased bone resorption in foetal mouse calvaria as a result of intermittent compressive forces *in vitro*. *Bone and Mineral* **2**, 441–8.

Klein-Nulend, J., Veldhuijzen, J. P., Van de Stadt, R. J., Van Kampen, G. P. J., Kuijer, R. & Burger, E. H. (1987b). Influence of intermittent compressive force on proteoglycan content of calcifying cartilage *in vitro*. *Journal of Biological Chemistry* **262**, 15490–5.

Lanyon, L. E. (1984). Functional strain as a determinant for bone remodelling. *Calcified Tissue International* **36**, S56–61.

Lanyon, L. E. (1987). Functional strain in bone tissue as an objective and controlling stimulus for adaptive bone remodelling. *Journal of Biomechanics* **20**, 1083–93.

Lozupone, E., Favia, A. & Grimaldi, A. (1992). Effect of intermittent mechanical force on bone tissue *in vitro*: preliminary results. *Journal of Bone and Mineral Research* **7**(S2), S407–9.

McDonald, F. & Houston, W. J. B. (1992). The effects of mechanical deformation on the distribution of potassium ions across the cell membrane of sutural cells. *Calcified Tissue International* **50**, 547–52.

McKeehan, W. L., Barnes, D., Reid, L., Stanbridge, E. Murakami, H. & Sato, G. H. (1990). Frontiers in mammalian cell culture *in vitro*. *Cellular and Developmental Biology* **26**, 9–23.

Maniotopoulos C., Sodek, J. & Melcher, A. (1988). Bone formation *in vitro* by stromal cells obtained from bone marrow of young adult rats. *Cell Tissue Research* **254**, 317–30.

Matsuda, J. & Davies, J. E. (1987). The *in vitro* response of osteoblasts to bioactive glass. *Biomaterials* **8**, 275–84.

Murray, D. W. & Rushton, N. (1990). The effect of strain on bone cell PGE2 release: a new experimental method. *Calcified Tissue International* **47**, 35–9.

Pead, M. J., Skerry, T. M. & Lanyon, L. E. (1988). Direct transformation from quiescence to formation in the adult periosteum following a single brief period of loading. *Journal of Bone and Mineral Research* **3**, 647–56.

Publicover, S., Thomas, G. & El Haj, A. J. (1994). Induction of a low-voltage activated fast inactivating Ca^{2+} channel in cultured bone marrow stromal cells by dex. *Calcified Tissue International* (in press).

Reich, K. M. & Frangos, J. A. (1991). Effect of flow on prostaglandin E2 and inositol triphosphate levels in osteoblasts. *American Journal of Physiology* **261**, C428–32.

Rubin, C. T. (1984). Skeletal strain and the functional significance of bone architecture. *Calcified Tissue International* **36**, S11–18.

Rubin, C. T. & Lanyon, L. E. (1984). Regulation of bone formation by applied dynamic loading. *Journal of Bone and Joint Surgery* **66A**, 397–402.

Shelton, R. M. & El Haj, A. J. (1991). Responses of rat periosteal and mouse MC3T3-E1 cells to PTH and PGE2 using a microcarrier bead column. *Journal of Physiology* **435**, 100 pp.

Shelton, R. M. & El Haj, A. J. (1992). A novel microcarrier bead model to investigate bone cell responses to mechanical compression *in vitro*. *Journal of Bone and Mineral Research* **7**, S403–7.

Wong, G. & Cohn, D. (1974). Separation of parathyroid hormone and calcitonin sensitive cells for nonresponsive bone cells. *Nature* **252**, 713–14.

Yeh, C. & Rodan, G. A. (1984). Tensile force enhances PGE2 synthesis in osteoblastic cells grown on collagen ribbons. *Calcified Tissue International* **36**, S67–71.

Zaman, G., Dallas, S. L. & Lanyon, L. E. (1992). Cultured embryonic bone shafts show osteogenic responses to mechanical loading. *Calcified Tissue International* **51**, 132–6.

T. M. SKERRY and B. F. FERMOR

Mechanical and hormonal influences *in vivo* cause regional differences in bone remodelling

Introduction

The purpose of this chapter is to show that the responses to osteotropic influences of different bones in the skeleton vary according to a 'survival value' associated with that particular bone. An understanding of the mechanisms by which these differences occur, particularly the ways in which some bones are resistant to resorption, could have pervasive consequences in the treatment of diseases which are characterised by functionally inappropriate bone loss.

Function of the skeleton

The skeleton has numerous functions, but the one which accounts predominantly for its shape, mass and material properties is the requirement that it should bear loads. Bone is unique among the tissues of the body in the level of its resistance to compressive forces, which results from its composition. While collagen and other organic components give tensile strength, hydroxyapatite is responsible for the resistance to compression. The relative amounts of these two components vary at different stages in life. Young animals have bones which are compliant as a result of a high proportion of organic components and relatively little mineral. This has been suggested to confer resistance to falls which would result in fractures in individuals with more rigid skeletons. Of course, the consequence of such compliance is greater dissipation of energy during loading. This translates into a structure which wastes more energy during locomotion than a more rigid bone (Currey, 1969). During maturation, mineralisation of the skeleton increases, and the individual usually becomes more coordinated, so that falls are less likely, and efficient

Society for Experimental Biology Seminar Series 54: *Biomechanics and Cells*, ed. F. Lyall & A. J. El Haj. © Cambridge University Press 1994, pp. 164–177.

locomotion is a more important requirement than high fracture threshold.

Regional differences in function and properties

The different bones of the skeleton have very different functions, and it would be simplistic to assume that all of them should be best served by one particular set of material properties. Examples at different ends of a spectrum of function are the antler of the deer, and the tympanic bulla (Currey, 1979). The antler is used in fighting, and must withstand high loads without failure, if the male is to succeed in acquiring a harem and passing on its genes. The tympanic bulla is a reference point for the sound sensing apparatus, yet in contrast to the antler, it rarely experiences any significant loads during life. The properties of the two bones reflect these functions. The antler is less mineralised than the bulla, and contains a higher proportion of organic components. Interestingly, neither of these bones is loaded in any significant way during their formation, so that their specific properties must be the result of perceived differences in cellular positional information during morphogenesis, rather than functional adaptation to use.

While these two examples are extremes, all of the bones of the skeleton vary in their individual functions and therefore in their mechanical requirements. Similarly, they vary in their importance to the individual's survival. The skull, for example, protects the brain. It is therefore essential that this bone should not be weakened, as the consequences of failure are severe. Ribs and some limb bones have lower survival value. While fracture of either would be incapacitating, it would not result inevitably in death.

The importance of strength of different bones in survival is borne out by comparison between the strains they normally experience, and those required to fracture them. In birds, the strains required to fracture the skull and the ulna are approximately similar. However, vigorous wing flapping generates 3000 microstrain in the ulna, while it is extremely difficult to induce the bird to perform any physiological activity which causes strains over 200 μstr in the skull (Lanyon, 1987; T. M. Skerry and R. A. Hillam, unpublished observations). Despite these differences, in adult animals, both these bones are relatively quiescent, and experience insignificant changes in mass in normal circumstances. If, however, the strains on the ulna are reduced to those normally experienced by the skull, intracortical and endosteal resorption occur (Rubin & Lanyon, 1987). It has been suggested that each bone has a genetically determined 'set point', below which it will not reduce its mass, however low its habitual strain exposure may fall (Lanyon, 1980). So in the skull, the set point would be higher than in the limbs, and disuse would have greater effects on those bones. The mechanism of this genetic set point or regional difference in susceptibility to the effects of osteotropic stimuli is clearly of great importance in the development of treatments for diseases such as osteoporosis.

Osteotropic influences – hormones and local factors

In order to discover the possible mechanisms for regional differences in bone cell actions, it is necessary to consider the osteotropic influences which are responsible for maintaining bone mass. Although bone remodelling has been shown experimentally to be profoundly influenced by the actions of many cytokines and eicosanoids (Goldring & Goldring, 1990; Saito et al., 1990; Das, 1991), it is difficult to specify the relevance of such findings in vivo. This is because, although the individual actions of osteotropic cytokines can be studied in vitro, in highly specified and characterised culture systems such experiments lead to numerous different and often contradictory results (Nathan & Sporn, 1991). This apparent paradox has been suggested to result from misconceptions in the consideration of targets for cytokine action. Instead of thinking of single cell types as cytokine targets, it may be more appropriate to see tissues as targets for their actions (Nathan & Sporn, 1991). This would allow the action of a given cytokine to be seen as the sum of its effects as a soluble, membrane or matrix-bound form, with effects on a total process within the tissue rather than only one aspect of a cell's metabolism.

After the cessation of growth, and in animals free from disease, the primary direct influences on bone cells are therefore those induced by the action of osteotropic hormones and mechanical loading. This does not imply that local cytokine or eicosanoid mediated actions are unimportant. However, they are more concerned with the flow of intercellular communication and the local regulation of the processes induced by the previous influences, than with the perception of more 'global' requirements to change bone mass.

The consequences of binding of some osteotropic hormones to cell surface receptors have been well characterised, while others are still fairly obscure. Parathyroid hormone (PTH), for example, has been shown to induce a series of intracellular changes, ranging from depolarisation of the cell, to influx and release of intracellular calcium, and G protein mediated cAMP expression (Herrman-Erlee, Lowik & Boonekamp, 1988; Davidson, Tatakis & Auerbach, 1990; Yellowley et al., 1993), which eventually impact on the cell's actions, and ultimately on bone mass. In contrast, the mechanisms of action of steroid hormones are less clear. 1,25-dihydroxyvitamin D (1,25D) has been shown to modulate genes for matrix molecules directly (Morrison *et al.*, 1989) as a result of the presence of a 1,25D response element in the relevant promoter region. Oestrogen, however, clearly acts to modulate bone mass, but by a mechanism which is currently poorly understood. The number of oestrogen receptors on bone cells is very low (Erikson *et al.*, 1988), and it is not universally accepted which cells are responsive to the hormone, or by what mechanism.

Osteotropic influences – mechanical changes

The mechanisms by which mechanical forces influence cell behaviour are similarly obscure. Neither has it been fully clarified which cells are the primary sensors of deformation of the matrix. Osteocytes are attractive in this role, and have been suggested by many workers to be more than cells trapped by advancing mineralisation (Lanyon, 1987; Van der Plas & Nijweide, 1992). Recent experimental work suggests that these cells are active in regulation of bone mass. Doty showed communication via gap junctions between osteocytes (Doty, 1981). In the absence of mechanical loading, the number and size of these gap junctional communications was reduced (Doty & Morey Holton, 1982).

More direct evidence for osteocyte responsiveness came from Lanyon's group, as a result of work on applied loading of avian bones *in vivo*. A single short period of mechanical loading, which if repeated daily would result in strain related increases in bone mass, was shown to have effects on osteocytes. Twenty-four hours after such a period of loading, in the absence of any significant intervening strains, osteocyte RNA expression (as measured by [³H]uridine uptake) was found to be increased approximately six-fold (Pead *et al.*, 1988). More remarkably still, it was shown that immediately after the same short (6 min) period of loading, there was a rapid strain-related increase in the activity of the enzyme glucose 6-phosphate dehydrogenase (G6PD) (Skerry *et al.*, 1989). A lack of any significant associated change in other enzymes studied (glyceraldehyde 3-phosphate dehydrogenase and aldolase) suggested that this G6PD activation was directed towards activation of the first part of the pentose monophosphate shunt only. This pathway forms the major source of the reducing equivalents necessary for biosynthesis, and of ribose sugars, necessary precursors for RNA synthesis.

However, it is evident that osteocytes are themselves incapable of performing significant changes in bone mass, despite the possibility of limited osteocytic resorption in the locale of the lacuna. Any major action must therefore be confined to signaling to osteoblasts on the surface, which are responsible for apposition of new bone, or for controlling osteoclastic resorption. These surface cells, whether they are active osteoblasts or quiescent lining cells, appear to be the key population in the control of bone mass. The interactions of mechanical and humoral influences must provide surface cells with sufficient information that they can initiate formation, resorption or simply maintain bone mass as appropriate to the prevailing requirements. Evidence for this is seen when changes in mechanical function, such as increases in load, which would normally be sufficient to increase bone formation, are able only to moderate resorption in the face of calcium deprivation (Lanyon, Rubin & Baust, 1986).

Regional differences in response

These interactions of mechanical and biochemical stimuli, which maintain bone strength, are influenced by some presumed genetically derived factor which relates to a bones function and position within the skeleton – the set point mentioned earlier. The examples previously cited related to specific bones within the skeleton, but the mechanism which determines these regional differences is likely to be more subtle than that. In a trabecular network within a bone, which is subjected to a resorptive stimulus, some elements will be preserved and others lost. This was suggested to be due to preservation of the elements loaded axially, and loss of the non-loaded ones. Such a view is simplistic, as compression of a complex structure of this sort is highly unlikely to result in such clear differences in the loading of different trabeculae. With a very few exceptions structures which are compressed become wider. This implies that lateral trabeculae will be subject to both axial and bending forces during such loading. The cells in the lateral and axial trabeculae (if such clearly demarcated groups were to exist) could be identical, and sensitive to strain in a way which is specific enough to distinguish between these two loading regimes. Alternatively, it may be the case that they have different set points for the perception of strains.

A further example is seen in all growing long bones. During growth, the diaphysis of a long bone usually apposes bone on the periosteal surface, and resorbs it from the endosteal surface. These diametrically opposite actions occur at either side of the cortex, which in a small animal may be only a few hundred micrometres thick. The surface cells from both endosteum and periosteum are phenotypically identical, have the same anastamotic blood supply, and are so close that any differences in perceived strains during loading are minimal. Despite these similarities, the actions of the endosteal and periosteal lining cells are opposite. A possible explanation for the mechanism by which this is achieved, but not its control is seen in Fig. 1 (Fermor *et al.*, 1993). TGFβ3 messenger RNA expression (as demonstrated by *in situ* hybridisation) occurs in two well demarcated zones of periosteal expression. No such expression is visible in endosteal cells, so that formation could be stimulated on one surface but not the other. Such specificity of localisation implies the same tight control of cell patterning and appreciation of positional information which has been seen in embryological development (Wolpert, 1977).

Regional differences in response of cells to osteotropic stimuli are not confined to response to mechanical changes (Wu *et al.*, 1990). In both experimental and clinical studies, such differences have been noted in response to hormonal changes. In post-menopausal osteoporotic patients, it is clear that some regions of the skeleton have an elevated risk of fracture owing to loss of bone matrix. Specific studies have shown that while the predilection sites for fracture – the femoral neck, vertebral body and distal radius – are subject to high loss, other regions are much less affected by changes in oestrogen status. The skulls of such individuals have been shown to be relatively insensitive to the reduction in circulating oestrogen which is associated with increased resorption in the other bones (Riggs *et al.*, 1981; Podenphant & Engel, 1987; Pun, Wong & Loh, 1991). Similarly, in stroke patients bone is preserved, despite bed rest which in other individuals would result in profound bone loss (Prince, Price & Ho, 1988). This suggests a role for the nervous system in the control of remodelling, which is hard to explain, given identical responses to loading in neurectomised and control bones *in vivo* (Hert, Liskova & Landa, 1971).

Experimentally, similar observations support these differences in effects of changes in oestrogen. In female rats, 4 weeks after ovariectomy, we found a 15% reduction in the thickness of the tibial cortex, compared with sham operated control animals (Skerry & Davy, 1993). In the same animals, there was no significant difference in the thickness of the calvariae (Fig. 2). Earlier changes were also detectable. Six days after ovariectomy, the uptake of an injected bolus of labelled calcium was reduced by 20% in the tibiae and vertebral bodies compared with controls (Skerry, 1992). As before, the calvariae of the two groups were not significantly different.

(a)

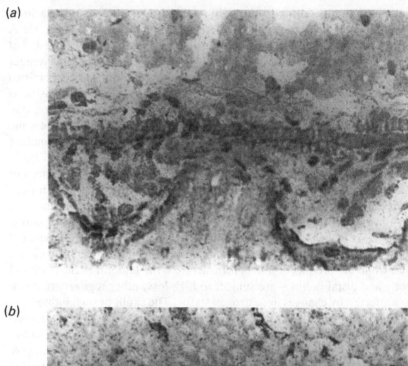

(b)

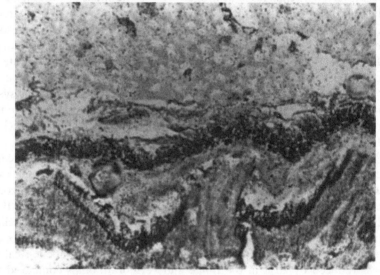

Fig. 1. TGFβ3 messenger RNA expression in rat bone counterstained with H & E. Sense control (a) shows low non-specific binding, while a serial section hybridised with antisense probe (b) shows two discrete bands of expression, in a cell layer remote from the periosteal surface, and also in the bottom of two regions undergoing active formation.

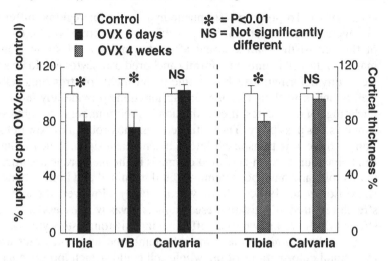

Fig. 2. Cortical bone thickness 4 weeks after ovariectomy (right), and acute calcium uptake 6 days after ovariectomy (left), in different regions of the skeleton. Ovariectomy had significant effects on the tibia, and 5th lumbar vertebral body (VB), but not on the calvariae.

Mechanisms for regional differences

The mechanisms of these differences are open to some speculation. Numerous cell culture experiments have demonstrated that bone-derived cells from calvariae and other parts of the skeleton have broadly similar phenotypes and responses to osteotropic hormones. Indeed, these properties are used as the criteria by which such cells are defined as bone cells. Since cells *in situ* have quite different responses to mechanical and hormonal stimuli, it may be that such criteria are flawed, as they actively select against cells which exhibit tissue specific properties, and towards those which are more uniform in their responses. Alternatively, the nature of the specificity may be connected with the cell's relationship with its matrix, not a cell when digested out of that environment. Such regulation would not be unique, as matrix interactions are capable of influencing cell behaviour in many other tissues. For example, it has been shown that basement membrane matrix regulates differentially the genes for TGFβ1, 2 and 3 in epithelial cells, by suppression or induction of the promoter regions (Streuli *et al.*, 1993).

Little is known about the nature of osteocyte matrix interactions, but it has been shown that the cell's lacunae are surrounded by proteoglycans which differ from those present throughout the rest of the matrix (Skerry *et al.*, 1990). To propose a simplistic hypothesis for regional differences in bone responses to strain, it is possible that the firmness of attachment of the cell to the matrix would affect transmission of strain from the matrix to the cell membrane. Firmly anchored osteocytes would be sensitive to tiny deformations which would be incapable of stretching cells less well attached to their lacunae. If gap junctions provide any mechanical anchorage of osteocytes, then reduction in their number in weightlessness supports this possibility. There still remains the problem of how deformation of the cell is transduced into a biochemical signal which influences cell behaviour. Stretch-activated channels in the membrane are attractive in this role, and have been demonstrated in animal and human bone cells (Davidson *et al.*, 1990; Yellowley *et al.*, 1993). However, the amount of stretch required to activate these channels is vastly in excess of the small strains which are known to be effective in activating osteocytes. If, however, focal attachments made the cell membrane sufficiently anisotropic, then small deformations of the whole cell could stretch loosely anchored areas considerably, amplifying stretching of those areas of membrane. Changes in these focal adhesions would therefore influence the transduction of forces, and have been observed in endothelial cells (Ingber, 1991).

The consequences of mechanical and hormonal stimulation of cells are therefore to produce a site-specific requirement for resorption, maintenance, or formation, which is an appropriate compromise between the demands of calcium homeostasis and mechanical function. The mechanism by which this interaction regulates cell behaviour is as poorly understood as the transduction mechanism. Logically, it is possible to hypothesise that different bones in the skeleton, or different regions of the same bone, are subject to the same circulating levels of osteotropic hormones. Local disuse or other processes which stimulate a requirement which is different from another area would appear to have two possible consequences. Either resorptive influences are associated with changes in number or affinity of cell surface receptors for agents stimulating resorption, or the consequences of binding of those agents are different in bone cells under different circumstances. In the first case, loading-related reduction in binding of a hormone such as PTH would achieve the required specificity, while in the second equal binding would elicit different second messenger responses. There is evidence to support differential expression and regulation of receptors in bone cells. In rats, it has been demonstrated that certain marrow cells designated parathyroid target (PT) cells bind hormone predominantly *in vivo*, although the studies failed to examine the binding of hormone to periosteal cells (Rouleau, Mitchell & Goltsman, 1990). These results contrast sharply with recent

(a)

(b)

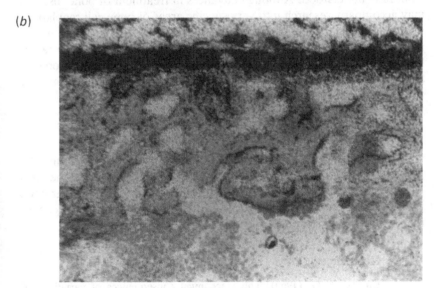

Fig. 3. PTH receptor messenger RNA expression in rat bone counterstained with H & E. Sense control (a) shows low non-specific binding, while the equivalent region in a serial section hybridised with antisense probe (b) shows intense expression in the periosteum, but not the endosteum. In addition, small focal areas of expression are visible on curved forming surfaces.

work (Fig. 3) which shows highly specific expression of PTH receptor messenger RNA in a band on the periosteum, and in trabecular lining cells, but not marrow cells (Fermor et al., 1993).

Even after ligand receptor binding, there are many levels at which regulation of action is possible. Immediate intracellular consequences of binding such as tyrosine kinase activation, and proto-oncogene expression are both complex processes which involve many steps dependent on other processes within the cell. Even if all these actions occur, it is possible to express messenger RNA which will not be transcribed into protein, or protein which will either not be secreted, or which will be inhibited before it can act. While many of these processes have been investigated in detail, there is no information on differences in their response to osteotropic influences in bone cells from different regions of the skeleton.

Conclusions

Clearly, the regulation of bone mass is a complex process, which is far from fully understood. Although advances in treatment of bone diseases may accompany research into very specific areas of bone cell regulation, the innate ability of some bone to resist the effects of stimuli which would initiate resorption in others represents an unexplored and exciting target for manipulation of bone loss. Given the incidence of clinical problems associated with inappropriate bone resorption, the potential rewards of advances in understanding of this phenomenon are great.

Acknowledgements

We would like to thank the Arthritis and Rheumatism Council and the Nuffield Foundation for funding. Dr Gino Segre kindly donated the PTH receptor probe, and Dr Anita Roberts the TGFβ3 probe. We are grateful to Dr D. Mason who was involved in the *in situ* hybridisation experiments, and Ms J. Harrington for photography.

References

Currey, J. D. (1969). The mechanical consequences of variations of mineral content of bone. *Journal of Biomechanics* 2, 1–11.

Currey, J. D. (1979). The mechanical properties of bones with greatly differing mechanical functions. *Journal of Biomechanics* 12, 313–19.

Das, U. N. (1991). Interaction(s) between essential fatty acids, eicosanoids, cytokines, growth factors and free radicals: relevance to new therapeutic strategies in rheumatoid arthritis and other collagen vascular diseases. *Prostaglandins Leukotrienes and Essential Fatty Acids* **44**, 201–10.

Davidson, R. M., Tatakis, D. W. & Auerbach, A. L. (1990). Multiple forms of mechanosensitive ion channels in osteoblast like cells. *Orthopedic Clinics of North America* **416**, 646–51.

Doty, S. B. (1981). Morphological evidence of gap junctions between bone cells. *Calcified Tissue International* **33**, 509–12.

Doty, S. B. & Morey Holton, E. R. (1982). Changes in osteoblastic activity due to simulated weightless conditions. *The Physiologist* **25**, S141–2.

Eriksen, E. F. Colvard, D. S., Berg, N. J., Graham, M. L., Mann, K. G., Spelsberg, T. C. & Riggs, B. L. (1988). Evidence of estrogen receptors in normal human osteoblast-like cells. *Science* **241**, 84–6.

Fermor, B., Mason, D. J., Lee, K., Segre, G. V., Gowen, M. & Skerry, T. M. (1993). *Journal of Bone and Mineral Research* **8**, S1–198.

Goldring, M. B. & Goldring, S. R. (1990). Skeletal tissue response to cytokines. *Clinical Orthopaedics and Related Research* **258**, 245–78.

Herrman-Erlee, M. P. M., Lowik, C. W. G. M. & Boonekamp, P. M. (1988). Different roles for calcium and cyclic AMP in the action of PTH: studies in bone explants and isolated bone cells. *Bone* **9**, 93–100.

Hert, J., Liskova, M. & Landa, J. (1971). Reaction of bone to mechanical stimuli. Part 5. Effect of intermittent stress on the tibia of the rabbit after resection of the peripheral nerves. *Folia Morphologica* **19**, 378–87.

Ingber, D. (1991). Integrins as mechanochemical transducers. *Current Opinion in Cell Biology* **3**, 841–8.

Lanyon, L. E. (1980). The influence of function on the development of bone curvature. An experimental study in the rat. *Journal of Zoology (London)* **192**, 457–66.

Lanyon, L. E. (1987). Functional strain in bone tissue as an objective, and controlling stimulus for adaptive bone remodelling. *Journal of Biomechanics* **20**, 1083–93.

Lanyon, L. E., Rubin, C. T. & Baust, G. (1986). Modulation of bone during calcium insufficiency by controlled dynamic loading. *Calcified Tissue International* **38**, 209–16.

Morrison, N. A., Shine, J., Fragonas, J.-C., Verkest, V., McEnemy, M. L. & Eisman, J. A. (1989). 1,25-Dihydroxyvitamin D-responsive element and glucocorticoid repression in the osteocalcin gene. *Science* **246**, 1158–61.

Nathan, C. F. & Sporn, M. B. (1991). Cytokines in context. *Journal of Cell Biology* **113**, 981–6.

Pead, M. J., Suswillo, R., Skerry, T. M., Vedi, S. & Lanyon, L. E. (1988). Increased ^{3}H-uridine levels in osteocytes following a single short period of dynamic bone loading *in vivo*. *Calcified Tissue International* **43**, 92–6.

Podenphant, J. & Engel, U. (1987). Regional variations in histomorphometric bone dynamics from the skeleton of an osteoporotic woman. *Calcified Tissue International* **40**, 184–8.

Prince, R. L., Price, R. I. & Ho, S. (1988). Forearm bone loss in hemiplegia: a model for the study of immobilisation osteoporosis. *Journal of Bone and Mineral Research* **3**, 305–10.

Pun, K. K., Wong, F. H. & Loh, T. (1991). Rapid postmenopausal loss of total body and regional bone mass in normal southern Chinese females in Hong Kong. *Osteoporosis International* **1**, 87–94.

Riggs, B. L., Wahner, H. W., Dunn, W. L., Mazess, R. B., Offord, K. P. & Melton, L. J. (1981). Differential changes in bone-mineral density of the appendicular and axial skeleton with aging – relationship to spinal osteoporosis. *Journal of Clinical Investigation* **67**, 328–35.

Rouleau, M. F., Mitchell, J. & Goltzman, D. (1990). Characterisation of the major parathyroid hormone target cell in the endosteal metaphysis of rat long bones. *Journal of Bone and Mineral Research* **5**, 1043–53.

Rubin, C. T. & Lanyon, L. E. (1987). Osteoregulatory nature of mechanical stimuli: function as a determinant for adaptive remodeling in bone. *Journal of Orthopaedic Research* **5**, 3001–10.

Saito, S., Ngan, P., Saito, M., Lanese, R., Shanfeld, J. & Davidovitch, Z. (1990). Interactive effects between cytokines on PGE production by human periodontal ligament fibroblasts *in vitro*. *Journal of Dental Research* **69**, 1456–62.

Skerry, T. M. (1992). Ovariectomy reduces uptake of calcium *in vivo* by the tibia and vertebral body but not the calvaria. *Calcified Tissue International* **50S**, 18.

Skerry, T. M., Bitensky, L., Chayen, J. & Lanyon, L. E. (1989). Early strain-related changes in enzyme activity in osteocytes following bone loading *in vivo*. *Journal of Bone and Mineral Research* **4**, 783–8.

Skerry, T. M. & Davy, A. (1993). Regional differences in bone remodelling in rats after ovariectomy. *Journal of Bone and Joint Surgery* (in press).

Skerry, T. M., Suswillo, R., El Haj, A. J., Ali, N. N., Dodds, R. A. & Lanyon, L. E. (1990). Load-induced proteoglycan orientation in bone tissue *in vivo* and *in vitro*. *Calcified Tissue International* **46**, 318–26.

Streuli, C. H., Schmidhauser, C., Kobrin, M., Bissell, M. J. & Derynck, R. (1993). Extracellular matrix regulates expression of the TGF-beta 1 gene. *Journal of Cell Biology* **120**, 253–60.

Van der Plas, A. & Nijweide, P. J. (1992). Isolation and purification of osteocytes. *Journal of Bone and Mineral Research* **4**, 389–96.

Wolpert, L. (1977). *The Development of Pattern and Form in Animals*. London: Oxford University Press.

Wu, D. D., Boyd, R. D., Fix, T. J. & Burr, D. B. (1990). Regional patterns of bone loss and altered bone remodeling in response to calcium deprivation in laboratory rabbits. *Calcified Tissue International* **47**, 18–23.

Yellowley, C., Levi, A. J., Skerry, T. M. & Hancox, J. (1993). Patch clamp recordings from human bone cells. *Journal of Physiology* (abstract in press).

L. E. LANYON

Mechanically sensitive cells in bone

Functionally adaptive control of bone architecture

Adaptation of bone architecture in relation to functional loading is a characteristic feature of bone physiology. In mammals and birds at least it is the mechanism on which the establishment and maintenance of load-bearing competence depends.

In connective tissues, unlike muscle, loads are sustained by an extracellular matrix. In tendon, ligament and cartilage resident cells throughout the structure may produce, remove and adapt matrix in their immediate vicinity, so regulating the amount of tissue present and altering its mechanical properties. In contrast, since bone matrix is rigid any change in bone architecture has to be accomplished by deposition of new tissue, and/or removal of old, either on existing surfaces or within voids excavated within the existing structure. The process of bone formation is performed by local mesenchymally derived osteoblasts whereas resorption requires haematogenously derived osteoclasts. Osteoclasts are, however, substantially influenced in their activity by osteoblasts (Chambers, 1985).

In bone, as in the other load-bearing connective tissues, one of the prime responsibilities of the resident cells is to maintain sufficient tissue present, with appropriate material properties, to sustain, without damage, the loads applied to it. Tissues subject to tension (such as tendon and ligament) self-align with load and so their shape is not critical. In contrast bones must withstand tension, compression, bending and torsion while providing the constant shape necessary for the close tolerances of joint surfaces and the consistent location of muscle attachments.

To maintain an appropriate architecture (including mass, geometry and material properties) the resident bone cells need information pertaining to the suitability of the existing architecture in relation to the

Society for Experimental Biology Seminar Series 54: *Biomechanics and Cells*, ed. F. Lyall & A. J. El Haj. © Cambridge University Press 1994, pp. 178–186.

loads applied. This information is most conveniently available through the strains produced as a result of functional load-bearing.

If strain is the controlling variable for this adaptive process it must at some stage be transduced into the biological language with which cells communicate and influence each others' behaviour.

Mechanotransduction is a property of many cell types (Watson, 1991) which has received considerable attention in bone cells. Although these tests have been carried out in osteoblast cultures where normal cell:cell and cell:matrix relationships are absent (i.e. Yeh & Rodan, 1984; Binderman *et al.*, 1988; Murray & Rushton, 1990; Jones & Bingmann, 1991), it is evident that strains of physiological magnitude do influence directly bone cell behaviour.

Mechanically related functional adaptation of bone architecture

In experiments in which the magnitude and/or distribution of normal functional strains has been altered by osteotomy of a supporting bone, apparently purposeful adaptive remodelling and renewed modelling occurs (i.e. Lanyon *et al.*, 1982; Burr *et al.*, 1989).

Some aspects of the adaptive response, such as the continued deposition of new bone on the cranial surface of the hypertrophying sheep radius despite strains on that surface being less than normal (Lanyon *et al.*, 1982), suggest that functional adaptation is not achieved simply through a direct influence of local strain magnitude on the cells responsible for formation and resorption on the bone's surfaces. Rather, it is likely to be controlled by a stimulus derived from an appreciation of strain distribution throughout the matrix (Lanyon, 1987).

Evidence from experiments where we have induced adaptive modelling/remodelling by direct control of bones' strain environment includes the following:

1. In the externally loadable adult avian ulna model *in vivo* it has been shown that bone mass can be substantially influenced in a peak strain magnitude-related manner by daily interruption of disuse by a short period of dynamic, but not static, loading (Lanyon & Rubin, 1984; Rubin & Lanyon, 1984, 1985). This influence on bone mass involves prevention of the resorption which normally occurs when loading is withdrawn, and a strain magnitude-related increase in periosteal and endosteal bone formation. Since substantial increases in bone area are produced in response to strains no larger than those naturally achieved by wing flapping, we postulate that the functional stimulus is derived from the difference in strain distribution between the two situations.

2. Exposure of the same avian preparation to a single period of such 'osteogenic' loading transforms the bone's quiescent periosteum to one actively forming bone 5 days later (Pead, Skerry & Lanyon, 1988a). It can be supposed that longer term adaptive modelling/remodelling responses are the cumulative effect of repeated exposure to such single stimuli. Five minutes after a single period of such loading the number of osteocytes showing activity of the enzyme glucose-6-phosphate dehydrogenase (G6PD) is increased by an amount that is related to the peak strain magnitude in their immediate locality (Skerry et al., 1989). The increased G6PD activity is not accompanied by an increase in the activity of glyceraldehyde 3-phosphate dehydrogenase or lactate dehydrogenase, indicating increased use of the pentose monophosphate shunt without any change in glycolysis. The pentose monophosphate pathway is involved in the synthesis of ribose sugars necessary for the production of RNA.

3. In rat tibiae subjected in vivo to a period of loading likely to be osteogenic, G6PD activity is similarly raised 5 min after loading in both osteocytes and surface cells (Dodds et al., 1993). In surface cells G6PD activity remains elevated for 24 h and is accompanied at that time by increased alkaline phosphatase activity (suggesting an osteogenic response at the bone's surface). In osteocytes 24 h after loading G6PD levels are restored to non-loaded levels. This suggests that osteocytes are more concerned with the earliest stages of the strain-related process, such as strain detection and the instigation of an adaptive response, than with subsequent control of adaptive modelling/remodelling. In osteocytes there is no ALP activity at any time. This is consistent with the absence of any matrix production by these cells.

4. In the avian model in vivo 24 h after a period of 'osteogenic' loading the number of osteocytes shown by autoradiography to be incorporating [³H]uridine is increased overall by a factor of 6 (12% to 72%) (Pead et al., 1988b). Surface cells also increase the level of their incorporation. Such changes are consistent with these cells increasing their RNA production. The distribution of cells showing increased [³H]uridine was not the same as that showing increased G6PD activity. The greatest increase in [³H]uridine uptake was instead in the cortex beneath the area of greatest subsequent new bone formation. This suggests possible 'signal processing' by the osteocyte population, transforming an initial (G6PD activity-based) response in resident cells which is directly proportional to the raw strain data, to a subsequent (RNA synthesis-dependent) controlling influence for the adaptive processes. The nature of the adaptive response may be related both to architecture and/or strain distribution.

5. Loading 17-day chick tibiae *in vitro* prevents the decline in alkaline phosphatase levels in periosteal cells associated with a period in culture, and increases the expression of the gene for collagen type 1 (Zaman, Dallas & Lanyon, 1992). This model therefore exhibits all the obligatory stages between strain change and enhanced osteogenic modelling activity. As in the adult avian situation *in vivo* this embryonic avian bone model shows a rapid, load magnitude-related elevation in G6PD activity in osteocytes followed over the subsequent 24 h by a strain magnitude-related increase in their RNA synthesis (Dallas *et al.*, 1993).

6. The osteogenic response to a single period of strain change in the adult avian model *in vivo* is substantially modified by a single high dose of indomethacin at the time of loading (Pead & Lanyon, 1989). In the embryonic chick tibia model *in vitro* the strain-related increase in both G6PD and RNA are also eliminated by indomethacin (Dallas *et al.*, 1993). This supports the hypothesis that adaptive modelling/remodelling requires an early prostanoid-dependent step which occurs prior to the increase in G6PD activity.

7. Mechanical loading of perfused cores of adult canine cancellous bone also shows loading-related increases in osteocytes' G6PD activity and RNA production (El Haj *et al.*, 1990; Rawlinson *et al.*, 1991) and allows demonstration of transient loading-related production of PGE and PGI_2. Immunocytochemistry confirms that the PGE is produced in surface lining cells but that the PGI_2 is present in both surface cells and osteocytes.

Perfusion of these cancellous cores with exogenous PGE_2 and PGI_2 shows that both prostanoids increase G6PD activity in both types of cell but that whereas PGI_2 mimics quantitatively the loading-related increase in RNA production there is no such response to PGE_2. PGI_2 but not PGE_2 also increases release of IGF-II, but not IGF-I, into the perfusing medium (Rawlinson *et al.*, 1993). Immunocytochemically both IGF-I and IGF-II can be identified in both osteoblasts and osteocytes.

8. Loading organ cultures of ulnae from 110 g rats, or treating them with exogenous prostanoids, confirms that PGI_2 and PGE_2 have different effects on bone cell behaviour. Prostacyclin and loading both increase G6PD activity, alkaline phosphatase activity and matrix synthesis. The level of stimulation of matrix synthesis is enhanced by oestrogen (Cheng, Zaman & Lanyon, 1993). PGE_2 also increases G6PD and matrix synthesis but has no effect on ALP activity. The influence of PGE_2 on matrix synthesis is not influenced by oestrogen.

It appears from this series of experiments that bone strain elicits more than one prostanoid response in osteoblasts and osteocytes, both of which may play a part in adaptive modelling and remodelling. Surface cells (which may themselves participate in matrix synthesis) are exposed directly to the effects of PGE, which clearly has an effect on osteogenesis (Jee *et al.*, 1985, 1990; Mori *et al.*, 1990) as well as prostacyclin. Osteocytes, which are not involved in matrix synthesis, are exposed only to the stimulus resulting from prostacyclin. The prime role of osteocytes in adaptive change in bone architecture must be regulatory and may be related to their communication of the effects of strain on themselves to those cells (predominantly on the surface) that are capable of changing bone architecture. Communication between osteocytes and surface osteoblasts is a feature of the bone cell population (Doty, 1981; Menton *et al.*, 1984; Shen, Kohler & Peck, 1986; Marotti *et al.*, 1990; Palumbo, Palazzinii & Marotti, 1990; Nefussi *et al.*, 1991).

Discussion and conclusions

The experimental results referred to here clearly indicate that bone modelling/remodelling can be profoundly influenced, in what seems a mechanically appropriate manner, by changes in the strain environment of the resident bone cells.

There is no convincing evidence on which aspect, or consequence, of strain change provides the actual strain-related influence on osteocytes *in vivo*. Ypey *et al.* (1992) are the only workers to our knowledge to have investigated the effects of strain on identified osteocytes and they suggest that stretch-activated ion channels are not involved. However, in addition to direct deformation of the cells themselves, dynamic strain inevitably involves changes in intra-lacunar pressure, fluid flow through the extracellular spaces, and the electrical potentials that flow of such charged fluid engenders. Any or all of these could directly affect osteocytes and surface osteoblasts in a strain-related manner.

It is perhaps relevant that the only cell population other than osteocytes which consistently produces prostacyclin is the endothelial cells lining the vascular tree. The surfaces of these cells are continuously subjected to the shearing forces of fluid flow and produce prostacyclin as a result (Frangos, Eskin & McIntire, 1985). Osteocytes are also subjected to intermittent shearing forces as fluid is forced by the deformation of the tissue through the canaliculi containing the osteocyte processes. Since fluid flow is strain-related and is capable of affecting bone cell behaviour it is possible that the actual loading-related variable to which bone cells respond *in vivo* is strain-induced fluid flow. This has been suggested previously in relation to PGE_2 (Reich & Frangos, 1991).

Since the fluid which is forced by strain past the cell bodies and processes of the osteocytes is electrically charged, fluid flow cannot be separated from the strain-generated electrical effects it produces. It has often been suggested that it is these electrical events which are the relevant strain-related variable in adaptive modelling/remodelling.

The hypothesis that osteocytes (and surface osteoblasts) are involved in the transduction of strain in bone tissue, and the subsequent influence of adaptive modelling/remodelling, has increasing experimental support. Osteocytes are rapidly responsive to mechanical events in their surrounding tissue in a peak strain magnitude-dependent manner. Their early responses include (in sequence) the production of prostacyclin, increased G6PD activity, increased RNA synthesis and possibly the production and/or release of IGF-II. Osteoblasts similarly respond with increased G6PD and alkaline phosphatase activity, increased matrix synthesis and possibly the production and/or release of IGF-II. This involvement of IGF-II is of particular interest since its production does not seem to be influenced by the traditional osteotrophic agents (McCarthy, Centrella & Canalis, 1992). This would be logical if its primary controlling influence were mechanical.

Exogenous administration of prostacyclin imitates all the early strain-related behaviour of osteocytes and osteoblasts which we have been able to demonstrate. Indomethacin at the time of loading eliminates or reduces all these responses.

The involvement of osteocytes and surface osteoblasts in (i) responding to strain (or one of its immediate consequences), (ii) forming an appreciation of strain distribution, and (iii) influencing modelling/remodelling, would account for the need for a population of live, strain-sensitive cells, distributed throughout the matrix and communicating with one another. This would not be necessary if loading-related adaptive modelling were controlled solely as a local response to strain in the immediate locality of cells on the bone surface.

Acknowledgements

Over the years the author has worked with many colleagues whose work is quoted here. This work has been supported by the Wellcome Trust, the MRC (UK), SERC, the Wolfson Foundation, the HRBLB, NASA and the USDA.

References

Binderman, I., Zor, U., Kaye, A. M., Shimshoni, Z., Harell, A. & Somjen, D. (1988). The transduction of mechanical force into biochemical events in bone cells may involve phospholipase A. *Calcified Tissue International* **42**, 261–6.

184   L. E. LANYON

Burr, D. B., Schaffler, M. B., Yang, K. H., Wu, D. D., Lukoschek, M., Kandzari, D., Sivaneri, N., Blaha, J. D. & Radin, E. L. (1989). The effects of altered strain environments on bone tissue kinetics. *Bone* **10**, 215–21.

Chambers, T. J. (1985). The pathology of the osteoclast. *Journal of Clinical Pathology* **38**, 241–52.

Cheng, Z. M., Zaman, G. & Lanyon, L. E. (1993). Estrogen enhances the stimulation of collagen synthesis by loading and exogenous prostacyclin, but not prostaglandin E_2, in organ cultures of rat ulnae. *Journal of Bone and Mineral Research* (in press).

Dallas, S., Zaman, G., Pead, M. J. & Lanyon, L. E. (1993). Early strain-related changes in bone cells in organ cultures of embryonic chick tibiae parallel those associated with adaptive modelling *in vivo*. *Journal of Bone and Mineral Research* **8**, 251–9.

Dodds, R. A., Ali, N., Pead, M. J. & Lanyon, L. E. (1993). Early loading-related changes in the activity of glucose 6-phosphate dehydrogenase and alkaline phosphatase in osteocytes and periosteal osteoblasts in rat fibulae *in vivo*. *Journal of Bone and Mineral Research* **8**, 261–7.

Doty, S. B. (1981). Morphological evidence of gap junctions between bone cells. *Calcified Tissue International* **33**, 509–12.

El Haj, A. J., Minter, S. L., Rawlinson, S. C. F., Suswillo, R. & Lanyon, L. E. (1990). Cellular responses to mechanical loading *in vitro*. *Journal of Bone and Mineral Research* **5**, 923–32.

Frangos, J. A., Eskin, S. G. & McIntire, L. V. (1985). Flow effects on prostacyclin production by cultured endothelial cells. *Science* **227**, 1477–9.

Jee, W. S. S., Mori, S., Li, X. I. & Chan, S. (1990). Prostaglandin E_2 enhances cortical bone mass and activates intracortical bone remodelling in intact and ovariectomized female rats. *Bone* **11.**, 253–66.

Jee, W. S. S., Ueno, K., Deng, Y. P. & Woodbury, D. M. (1985). The effects of prostaglandin E_2 in growing rats: increased metaphyseal hard tissue and cortico-endosteal bone formation. *Calcified Tissue International* **37**, 148–57.

Jones, D. B. & Bingmann, D. (1991). How do osteoblasts respond to mechanical stimulation? *Cells and Materials* **1**, 1–12.

Lanyon, L. E. (1987). Functional strain in bone tissue as an objective and controlling stimulus for adaptive bone remodelling. *Journal of Biomechanics* **20**, 1083–93.

Lanyon, L. E., Goodship, A. E., Pye, C. J. & MacFie, J. H. (1982). Mechanically adaptive bone remodelling. *Journal of Biomechanics* **15**, 141–54.

Lanyon, L. E. & Rubin, C. T. (1984). Static versus dynamic loads an influence on bone remodelling. *Journal of Biomechanics* **17**, 897–905.

McCarthy, T. L., Centrella, M. & Canalis, E. (1992). Constitutive synthesis of insulin-like growth factor-II by primary osteoblast-enriched cultures from fetal rat calvariae. *Endocrinology* **130**, 1303–8.

Marotti, G., Cane, V., Palazzini, S. & Palumbo, C. (1990). Structure–function relationships in the osteocyte. *Italian Journal of Mineral and Electrolyte Metabolism* **4**, 93–106.

Menton, D. N., Simmons, D. J., Chang, S. A.-L. & Orr, B. Y. (1984). From bone lining cell to osteocyte – an SEM study. *Anatomical Record* **209**, 29–39.

Mori, S., Jee, W. S. S., Li, X. J., Chan, S. & Kimmel, D. B. (1990). Effects of prostaglandin E₂ on production of new cancellous bone in the axial skeleton of ovariectomized rats. *Bone* **11**, 103–13.

Murray, D. W. & Rushton, N. (1990). The effect of strain on bone cell prostaglandin release: a new experimental method. *Calcified Tissue International* **47**, 35–9.

Nefussi, J. R., Sautier, J. M., Nicholas, V. & Forest, N. (1991). How osteoblasts become osteocytes; a decreasing matrix forming process. *Journal de Biologie Buccale* **19**, 75–82.

Palumbo, C., Palazzinii, S. & Marotti, G. (1990). Morphological study of intercellular junctions during osteocyte differentiation. *Bone* **11**, 401–6.

Pead, M. J. & Lanyon, L. E. (1989). Indomethacin modulation of load-related stimulation of new bone formation *in vivo*. *Calcified Tissue International* **45**, 34–40.

Pead, M. J., Skerry, T. M. & Lanyon, L. E. (1988a). Direct transformation from quiescence to bone formation in the adult periosteum following a single brief period of bone loading. *Journal of Bone and Mineral Research* **3**, 647–56.

Pead, M. J., Suswillo, R., Skerry, T. M., Vedi, S. & Lanyon, L. E. (1988b). Increased ³H uridine levels in osteocytes following a single short period of dynamic bone loading *in vivo*. *Calcified Tissue International* **43**, 92–6.

Rawlinson, S. C. F., El Haj, A. J., Minter, S. L., Tavares, I. A., Bennett, A. & Lanyon, L. E. (1991). Loading-related increases in prostaglandin production in cores of adult canine cancellous bone *in vitro*: a role for prostacyclin in adaptive bone remodelling? *Journal of Bone and Mineral Research* **6**, 1345–51.

Rawlinson, S. C. F., Mohan, S., Baylink, D. J. & Lanyon, L. E. (1993). Exogenous prostacyclin, but not prostaglandin E₂, produces similar responses in both G6PD activity and RNA production as mechanical loading, and increases IGF-II release in adult cancellous bone in culture. *Calcified Tissue International* **53**, 324–9.

Reich, K. M. & Frangos, J. A. (1991). Effect of flow on prostaglandin E₂ and inositol trisphosphate levels in osteoblasts. *American Journal of Physiology*, C428–32.

Rubin, C. T. & Lanyon, L. E. (1985). Regulation of bone mass by mechanical strain magnitude. *Calcified Tissue International* **37**, 411–17.

Rubin, C. T. & Lanyon, L. E. (1984). Regulation of bone mass by applied dynamic loads. *Journal of Bone and Joint Surgery* **66A**, 397–402.

Shen, V., Kohler, G. & Peck, W. A. (1986). Prostaglandins change cell shape and increase intercellular gap junctions in osteoblasts cultured from rat fetal calvaria. *Journal of Bone and Mineral Research* **1**, 243–9.

Skerry, T. M., Bitensky, L., Chayen, J. & Lanyon, L. E. (1989). Early strain-related changes in enzyme activity in osteocytes following bone loading *in vivo*. *Journal of Bone and Mineral Research* **4**, 783–8.

Watson, P. A. (1991). Function follows form: generation of intracellular signals by cell deformation. *FASEB Journal* **5**, 2013–19.

Yeh, C.-K. & Rodan, G. A. (1984). Tensile forces enhance prostaglandin E synthesis in osteoblastic cells grown on collagen ribbons. *Calcified Tissue International* **36**, S67–71.

Ypey, D. L., Weidema, A. F., Hold, K., van der Laarne, A., Ravesloaf, J. H., van der Plas, A. & Nijweide, P. J. (1992). Voltage, calcium and stretch activated ionic channels and intracellular calcium in bone cells. *Journal of Bone and Mineral Research* **7**, S377–87.

Zaman, G., Dallas, S. L. & Lanyon, L. E. (1992). Cultured embryonic shafts *in vitro* show osteogenic responses to loading. *Calcified Tissue International* **51**, 132–6.

E. H. BURGER, J. KLEIN-NULEND
and J. P. VELDHUIJZEN

Mechanical stress and bone development

Introduction

It has been well established that the skeleton, just like the muscular
apparatus, responds to changes in applied mechanical stress by changing
its mass. This phenomenon was first described by Wolff some hundred
years ago in his book *Das Gesetz der Transformation der Knochen*
(Wolff, 1986). Wolff was struck by gross changes in trabecular organisation of bones as a result of dramatic changes of their mechanical environment, as for instance occurs after maladjustment of the bone ends
after a fracture. Wolff's Law, which is in fact not a law but the description
of a biological phenomenon, states that bones adapt their tissue density
and architecture to the functional demands of their mechanical environment, to obtain maximal strength with minimal tissue mass. Increased
stress leads to denser bone, as in the dominant arm of professional tennis
players (Montoye, Smith & Fardon, 1980), while stress reduction, as
for instance after long-term bed rest, reduces bone density (Whedon &
Heaney, 1993).

While these phenomena are quite apparent at the anatomical and
tissue level, it is not well known how they are achieved at the cellular
and molecular level. How skeletal cells react to mechanical stress in
terms of cell proliferation and matrix metabolism, and what intermediate
molecules are involved, are still matters of debate. Even the nature of
the final physical stimulus which acts on the cell to modulate its behaviour
has not been unequivocally determined.

On the other hand, several methods have now been developed which
allow study of cellular reactions to some form of strain, either by pulling
the substrate on which the cells grow, applying a flow of fluid over the
cells, or submitting them to compressive hydrostatic force. Skeletal cells

Society for Experimental Biology Seminar Series 54: *Biomechanics and Cells*, ed. F. Lyall &
A. J. El Haj. © Cambridge University Press 1994, pp. 187–196.

respond to such treatment by a variety of reactions (Veldhuijzen, Bour-ret & Rodan, 1979; Murray & Rushton, 1990; Reich, Gay & Frangos, 1990). However, the cells used are mainly derived from bones of embryonic, fetal or neonatal animals. The susceptibility of such rapidly growing bones to mechanical stimuli may be questioned, because the role of mechanical stress during skeletal morphogenesis has not been well established. Several authors consider that functional adaptation is a phenomenon of the adult skeleton, while bone development is guided by hormonal regulation only (Carter et al., 1987). In the following pages a series of experiments is reviewed in which whole embryonic mouse bones were subjected to intermittent hydrostatic compression in vitro, during organ culture. These studies provide strong evidence that embryonic bones are indeed sensitive to mechanical stress. Matrix metabolism and cell differentiation are particularly sensitive to the mechanical environment. This makes such bone organ cultures useful for the study of skeletal adaptation at the cellular level, and indicates that all differentiated skeletal cells are mechanosensitive, independent of the age of the individual.

Application of intermittent compressive force (ICF) to developing bone rudiments

The technique of applying hydrostatic pressure pulses to bone organ cultures has been described elsewhere (Klein-Nulend, Veldhuijzen & Burger, 1986; Burger, Klein-Nulend & Veldhuijzen, 1991). Briefly, bones are kept in tissue culture dishes placed in a closed culture chamber (98% humidity) in an incubator at 37 °C. The gas phase (5% CO_2 in air) was intermittently compressed using a compression cylinder attached to the culture chamber (Fig. 1). Pulse frequency was 0.3 Hz, 1 s pressure followed by 2 s relaxation, and pressure maximum was reached within 400 milliseconds.

The bone organs cultured were long bones from the embryonic mouse metatarsus or half calvariae. In some cases radii and ulnae were also used. The bones were derived from 15-, 16- or 17-day-old mouse embryos, depending on the parameter studied.

In 15-day-old mouse embryos, the metatarsal bone rudiments consist entirely of primitive cartilage and ossification has not yet started. At 16 days of embryonic life, cartilage hypertrophy is apparent in the rudiment centre and at histological examination a single layer of osteoblasts is found in the periosteum lining this hypertrophic zone. The osteoblasts have produced a thin sheath of osteoid, but calcification has not yet started. One day later, at 17 days of gestation, calcification has started in the hypertrophic cartilage centre as well as in the bone layer sur-

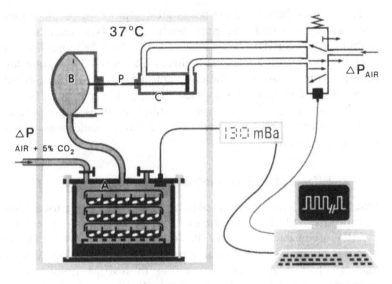

Fig. 1. Diagram of the apparatus for application of intermittent compressive force (ICF). Culture dishes are placed in culture vessel A which is connected with cylinder B. The gas phase (5% CO_2 in air) of A and B is compressed by moving piston P. P is moved by inlet of compressed air, alternatingly in one of 2 ports of cylinder C. A, B and C are kept at 37 °C in an incubator.

rounding it. In histological sections stained for tartrate-resistant acid phosphatase (TRAP), mononuclear TRAP-positive cells are now found between the osteoblasts, and at $17\frac{1}{2}$ days they are beginning to migrate towards the bone's centre. They pass the (thin) sheath of mineralised bone, which has been laid down in a trabecular fashion, leaving several uncalcified areas which can be easily traversed by any type of periosteal cell. While migrating into the bone rudiment the TRAP cells fuse, and develop into mature osteoclasts. These cells start to resorb the network of mineralised cartilage matrix of the bone's centre, thereby excavating the primitive marrow cavity. Endothelial sprouts follow in their wake, as well as fibroblast-like cells which set up the primitive marrow stroma filling the cavity which once held calcified cartilage. At 18 days of embryonic age the marrow cavity has been established, the bone collar of the diaphysis has thickened and now consists of two layers of trabecular bone and the various steps of chondrocyte proliferation–chondrocyte flattening–chondrocyte hypertrophy and matrix calcification occur at both sides (proximal and distal) of the primitive marrow cavity.

Throughout this period the bones grow rapidly, by cartilage proliferation in the non-osteogenic ends, leading to a 300% increase in length within 3 days. Similar events take place in the other long bones of the skeleton, but mostly somewhat earlier, while the phalanges develop even later than the metatarsal rudiments. This means that 17-day-old radii and ulnae are at a stage of development comparable to a late-18-day metatarsal bone. In addition, the outer metatarsal bones, supporting digits 1 and 5, are about half a day later than the three middle ones, which were used in our studies.

Effects of ICF on matrix mineralisation

Our first studies on the effects of pulsed hydrostatic force used 16-day-old metatarsals, and we followed mineralisation of the bone's centre during a 5-day organ culture period (Klein-Nulend et al., 1986). The culture medium originally consisted of a complicated mixture of chick embryo extract, chick plasma and rat serum in αMEM (minimal essential medium), but in later studies similar results of pressure were found in a medium consisting solely of αMEM with 0.1% bovine serum albumin (Bagi & Burger, 1989). Pressure height was usually 13 kPa, or 130 mbar above ambient pressure. We calculated (Klein-Nulend et al., 1986) that in the intact embryo, the maximal pressure put on the metatarsal rudiments as a result of contraction of all the relevant muscles in the embryonic leg is about 12 kPa.

In organ culture, embryonic bones generally grow rapidly even in the absence of serum. Chondrocyte hypertrophy and osteoblast development also occur fairly rapidly, at about half the speed observed in vivo. However, matrix mineralisation as a rule is severely retarded, leading to wide zones of non-calcified hypertrophic cartilage on both sides of a very short mineralised centre, while the bone collar often hardly mineralises at all.

Culture under 13 kPa ICF accelerated calcification, leading to a calcified zone twice as long after 5 days of culture compared with control cultures (Fig. 2), or a 240% increased uptake of ^{45}Ca from the medium (Klein-Nulend et al., 1986). Phosphate uptake was also enhanced, and the Ca/P ratio of the newly formed mineral did not differ from control cultures. Alkaline phosphatase, an enzyme closely linked to matrix mineralisation was enhanced some 30% after 5 days compared with control cultures. Histologically the bone rudiments were healthy, and no adverse effects of ICF treatment were observed (Klein-Nulend et al., 1986).

In experiments where the pressure applied in vitro was lowered, we found that 10 kPa (100 mbar) still caused an almost full-blown effect, while 5 kPa (50 mbar) had no effect at all (Fig. 3). This indicates that

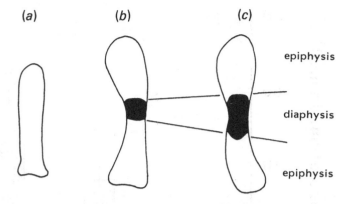

(a) (b) (c)

epiphysis

diaphysis

epiphysis

Fig. 2. Line drawings of metatarsal rudiments, before and after culture. (*a*) non-cultured, uncalcified rudiment; (*b*) control rudiment after 5 days' culture. In the centre, a dark spot has appeared as a result of matrix mineralisation; (*c*) experimental rudiment, after 5 days' culture under 130 mbar ICF. The diaphyseal area of calcified matrix is about twice as long as in the paired control culture.

there is a threshold pressure below which no effect occurs, similar to the study of Veldhuijzen *et al.* (1979) on chondrocyte proliferation.

The stimulating effect on alkaline phosphatase activity, together with the finding that ICF had no effects in bones which were killed by freezing and thawing before culture (Klein-Nulend *et al.*, 1986), showed that intermittent compression stimulated cellular processes related to mineralisation and did not act on the physical process of mineralisation. Also, the effects could not be mimicked by increasing the partial pressure of CO_2 and O_2 by 13% (5% $CO_2 \rightarrow 5.6\%$, 20% $O_2 \rightarrow 23.2\%$), which meant that ICF did not work via a change in the gas phase composition (Klein-Nulend *et al.*, 1986). Increased CO_2 concentration would anyway lead to a lower pH, which favours mineral resorption rather than mineral deposition.

However, if ICF acted via mechanical perturbations of the tissue, how should this be described in terms of stress and strain? This is a difficult question, because the mechanical properties of embryonic mouse cartilage and bone have not yet been reported. As a result only the distribution of stress within the bone rudiment can be calculated, using finite element analysis of the embryonic bones. Such an analysis was performed by Wong & Carter (1990), who showed that externally applied hydrostatic pressure produces significant shear stresses at the cartilage/calcified cartilage interface as a result of material mismatch, and pure hydrostatic

192   E. H. BURGER, J. KLEIN-NULEND & J. P. VELDHUIJZEN

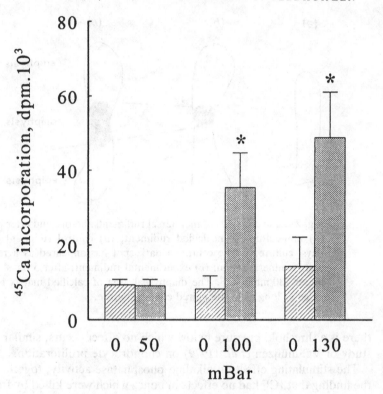

Fig. 3. Effect of 5 days culture under 50, 100 or 130 mbar ICF on calcium incorporation in uncalcified metatarsal bone rudiments. Bones were cultured for 5 days in medium containing 1 µCi ^{45}CaCO$_3$ (sp. act. 1 mCi mmol^{-1}) per ml. After 5 days, incorporation of radiolabelled ^{45}Ca into the mineral phase was determined by decalcification of the rudiments in 1 ml 5% formic acid for 24 h, followed by liquid scintillation counting. Values are $x \pm$ SEM, $n = 8$ paired bones. * $p < 0.01$, Student's t-test.

pressure at the homogeneous rudiment ends. The accelerated osteogenesis in regions of high shear and the increased synthesis of proteoglycans in regions of high compressive hydrostatic stress were therefore consistent with a theory developed by Carter (1987). This theory predicts that the replacement of cartilage by bone during endochrondral ossification is accelerated by intermittent shear stress. Hydrostatic stress, on the other hand, would favour cartilage matrix formation. Indeed, in the rudiment ends ICF increased the synthesis of cartilage proteoglycans, but not in the calcifying centre (Klein-Nulend et al., 1987a). In addition, ICF treatment accelerates mineralisation of the rudiments only after material heterogeneity has been established, as a result of cartilage hypertrophy in the rudiment centre (Burger, Klein-Nulend & Veldhuijzen, 1992). This data corroborates the conclusion of Wong & Carter (1990) that shear stress, not hydrostatic stress, accelerates endochondral ossification. However, for determination of the amount of shear *strain* applied to the cultured bones, data about the material behaviour of their components (embryonic cartilage and embryonic bone) are needed, such as Young's modulus. We are currently attempting to obtain these data from pieces of embryonic mouse rib. Preliminary data (E. H. Burger, R. Huiskes, S Goldstein and K. Jepsen, unpublished results) indicate that the elastic modulus of embryonic cartilage is of the order of 2 mPa, while that of embryonic bone is about two orders of magnitude higher. Combining these data with those of Wong and Carter (1990), 13 kPa hydrostatic pressure produces around 1000 microstrain at the interface between calcified and non-calcified tissues in the embryonic rudiment.

Cell-signal molecules produced by rudiments under ICF

The signal transduction pathway and the intermediate molecules involved in the response to ICF have not yet been studied in depth. However, a few studies have been published, which indicate that bone organs under ICF produce soluble factors capable of inducing anabolic responses in bone cells. In addition, osteoclast formation and mineral resorption are inhibited.

To start with the latter: in 17-day-old embryonic metatarsal rudiments, culture under ICF inhibits the influx of osteoclastic cells into the calcified bone centre, formation of a marrow cavity and release of ^{45}Ca from pre-labelled bones (Klein-Nulend *et al.*, 1990). Apparently ICF interferes with the processes of osteoclast recruitment and chemotaxis which take place under organ culture conditions as well as *in vivo* (Dieudonné *et al.*, 1991).

In addition, conditioned media from bones cultured under ICF also inhibited the release of ^{45}Ca from fresh, untreated bones indicating that one or more soluble factors were involved (Klein-Nulend *et al.*, 1990). In cultures of adult mouse bone marrow ICF treatment reduced the formation of cells positive for tartrate-resistant acid phosphatase (TRAP), a marker of osteoclastic cells (Van Wijngaert, Tas & Burger, 1987). Again, TRAP-cell formation was also inhibited by the conditioned media of ICF-treated bones (Klein-Nulend *et al.*, 1990). In calvariae of 17-day-old embryonic mice, 4–5 days of ICF treatment caused anabolic effects on mineral metabolism. As in the long bones, bone mineralisation

194   E. H. BURGER, J. KLEIN-NULEND & J. P. VELDHUIJZEN

was stimulated and alkaline phosphatase activity enhanced (Klein-Nulend *et al.*, 1987b). In addition, ICF caused an increased DNA synthesis in the bones (Klein-Nulend *et al.*, 1993) and inhibited osteoclastic resorption, measured as release of calcium (Klein-Nulend *et al.*, 1987b). Conditioned media of ICF-treated calvariae increased DNA synthesis in fresh calvariae, an effect which was followed after a few days by enhanced synthesis of collagen but not non-collagenous proteins. Also, ICF-conditioned media inhibited TRAP-cell formation in bone marrow cultures, as did media from long bones (Klein-Nulend *et al.*, 1993). Apparently, ICF caused the calvariae to release autocrine growth factors stimulating bone formation as well as paracrine factor inhibiting osteoclast formation.

Although these studies strongly suggest that (some of) the effects of ICF are mediated by soluble factors released from the cells, the identity of these factors has not yet been established. Some of their effects are mimicked by exogenous PGE_2, such as the stimulation of DNA synthesis followed by collagen production in the calvariae (Raisz & Kream, 1983). Also, the inhibiting effects on osteoclast formation and their influx into the bone rudiment are mimicked by $TGF\beta_1$ (Dieudonné *et al.*, 1991). Both these signal molecules, $TGF\beta$ and PGE_2, have been implicated in the response of bone tissue to mechanical stress (Bindermann *et al.*, 1988; Turner, 1992). Both molecules are also produced by bone organ cultures *in vitro* (Klein-Nulend *et al.*, 1991; Dieudonné *et al.*, 1993). However, bone rudiments produce many different growth factors and cytokines and the exact identity of the factor(s) mediating the effects of ICF on intact bone remains to be established.

In conclusion, there is good evidence that mild mechanical stress as a result of pulsed hydrostatic compression at near-physiological magnitude has many effects on the behaviour of the cells in organ cultures of bone. Bone formation appears to be stimulated, while bone resorption is inhibited by this treatment. Considering that bones under organ culture conditions are deprived of all mechanical stress except gravity, the lack of mineralisation may now be interpreted as resulting from a lack of mechanical stimulation, which *in vivo* occurs as a result of muscle contractions in the legs and feet of the embryo (Vries, Visser & Prechtl, 1982). ICF seems to restore in a crude manner this 'mechanical environment', and the anabolic effect of such treatment on bone formation provides good evidence that indeed even embryonic bones are sensitive to mechanical stress and strain. Apparently all differentiated cells of cartilage and bone are mechanosensitive, independent of the stage of differentiation and age of the intact organism.

References

Bagi, C. & Burger, E. H. (1989). Mechanical stimulation by intermittent compression stimulates surface incorporation and matrix mineralization in fetal mouse long-bone rudiments under serum-free conditions. *Calcified Tissue International* **45**, 342–7.

Binderman, I., Zor, U., Kaye, A. M., Shimshoni, Z., Harell, A. & Somjen, D. (1988). The transduction of mechanical force into biochemical events in bone cells may involve activation of phospholipase A2. *Calcified Tissue International* **42**, 261–6.

Burger, E. H., Klein-Nulend, J. & Veldhuijzen, J. P. (1991). Modulation of osteogenesis in fetal bone rudiments by mechanical stress *in vitro*. *Journal of Biomechanics* **24**(S1), 101–9.

Burger, E. H., Klein-Nulend, J. & Veldhuijzen, J. P. (1992). Mechanical stress and osteogenesis *in vitro*. *Journal of Bone and Mineral Research* **7** S397–401.

Carter, D. R. (1987). Mechanical loading history and skeletal biology. *Journal of Biomechanics* **20**, 1095–109.

Carter, D. R., Orr, T. E., Fyhrie D. P. & Schurman, D. J. (1987). Influences of mechanical stress on prenatal and postnatal skeletal development. *Clinical Orthopaedics* **219**, 237–50.

Dieudonné, S. C., Foo, P., Van Zoelen, J. & Burger, E. H. (1991). Inhibiting and stimulating effects of TGF-β₁ on osteoclastic bone resorption in fetal mouse bone organ cultures. *Journal of Bone and Mineral Research* **6**, 479–87.

Dieudonné, S. C., Semeins, C. M., Goei, S. W., Vukicevic S., Klein-Nulend, J., Sampath, T. K., Helder, M. & Burger, E. H. (1993). Opposite effects of Transforming Growth Factor beta and osteogenic protein on chondrogenesis in cultured long bone rudiments. *Journal of Bone and Mineral Research* (in press).

Klein-Nulend, J., Veldhuijzen, J. P. & Burger, E. H. (1986). Increased calcification of growth plate cartilage as a result of compressive force *in vitro*. *Arthritis and Rheumatism* **29**, 1002–9.

Klein-Nulend, J., Veldhuijzen, J. P., Van de Stadt, R. J., Van Kampen, G. P. J., Kiujer, R. & Burger, E. H. (1987a). Influence of intermittent compressive force on proteoglycan content of calcifying growth plate cartilage *in vitro*. *Journal of Biological Chemistry* **262**, 15490–5.

Klein-Nulend, J., Veldhuijzen, J. P., De Jong, M. & Burger, E. H. (1987b). Increased bone formation and decreased bone resorption in fetal mouse calvaria as a result of intermittent compressive force *in vitro*. *Bone and Mineral* **2**, 441–8.

Klein-Nulend, J., Veldhuijzen, J. P., Van Strien, M. E., De Jong, M. & Burger, E. H. (1990). Inhibition of osteoclastic bone resorption by mechanical stimulation *in vitro*. *Arthritis and Rheumatism* **33**, 66–72.

Klein-Nulend, J., Aaron, J. N., Harrison, J. R., Simmons, H. A. & Raisz, L. G. (1991).Mechanism of regulation of prostaglandin production by parathyroid hormone, interleukin-1, and cortisol in cultured mouse parietal bones. *Endocrinology* **128**, 2503–10.

Klein-Nulend, J., Semeins, C. M., Veldhuijzen, J. P. & Burger, E. H. (1993). Effect of mechanical stimulation on the production of soluble bone factors in cultured fetal mouse calvariae. *Cell and Tissue Research* **271**, 513–17.

Montoye, H. J., Smith, E. L. & Fardon, D. F. (1980). Bone mineral in senior tennis players. *Scandinavian Journal of Sports Science* **2**, 26–32.

Murray, D. W. & Rushton, N. (1990). The effect of strain on bone cell prostaglandin E_2 release: a new experimental model. *Calcified Tissue International* **47**, 35–9.

Raisz, L. G. & Kream, B. E. (1983). Regulation of bone formation. *New England Journal of Medicine* **309**, 29–35.

Reich, K. M., Gay, C. V. & Frangos, J. A. (1990). Fluid shear stress as a mediator of osteoblast cyclic adenosine monophosphate production. *Journal of Cell Physiology* **143**, 1001–4.

Turner, C. H. (1992). Functional determinants of bone structure: beyond Wolff's Law of bone transformation. *Bone* **13**, 403–9.

Van Wijngaert, F. P., Tas, M. C. & Burger, E. H. (1987). Characteristics of osteoclast precursor-like cells grown from mouse bone marrow. *Bone and Mineral* **3**, 111–23.

Veldhuijzen, J. P., Bourret, L. A. & Rodan, G. A. (1979). *In vitro* studies of the effect of intermittent compressive forces on cartilage cell proliferation. *Journal of Cell Physiology* **98**, 299–306.

Vries, J. I. P. de, Visser, G. H. A. & Prechtl, H. F. R. (1982). The emergence of fetal behaviour. I. Qualitative aspects. *Early Human Development* **7**, 301–22.

Whedon, C. D. & Heaney, R. P. (1993). Effects of physiological activity paralysis and weightlessness on bone growth. In *Bone*, Vol. 7, *Bone Growth B*, ed. B. K. Hall, pp. 57–77. Boca Raton: CRC Press.

Wolff, J. (1986). *The law of bone remodeling (Das Gesetz der Transformation der Knachen, 1892)*. Translated by P. Magnet & R. Furlong. Berlin: Springer.

Wong, M. & Carter, D. R. (1990). Theoretical stress analysis of organ culture osteogenesis. *Bone* **11**, 127–31.

D. B. JONES, G. LEIVSETH, Y. SAWADA,
J. VAN DER SLOTEN and D. BINGMANN

Application of homogenous, defined strains to cell cultures

Introduction

Nearly all types of cells, prokaryotic and eukaryotic, can respond to the mechanical environment. This environment may be due to environmental stresses (e.g. wind, or activity, perhaps also gravity) or vibrations (e.g. sound), the difference being merely the amplitudes and the frequencies involved. Distortions of some parts of the cell or tissue are caused and this distortion is transduced into an electrophysiological and/or biochemical response. In some protists, organelles have become adapted into vibration sensors and can elicit evasion or aggression responses. In many multicellular organisms groups of cells can be adapted as vibration sensing-organs. For instance, in coelenterates special cells act as sensors to release a trigger mechanism controlling the nematocysts for prey capture (Golz & Thurm, 1991). Plants can adapt to environmental wind forces and also to gravity (Gehring *et al.*, 1990). Specialised organs exist in animals to transduce small mechanical loads, such as vibrations, and link this transducer to the nervous system. Strains induced by stresses in the swim bladder of fish are detected, which can aid in prey capture or predator evasion (Canfield & Eaton, 1990). The sensory hairs of vestibular ampullae (similar to the cochlea) provide an example of an organ where small vibrations are amplified by exploiting the production of shear between two surfaces (Nagel *et al.*, 1991). Interaction with the mechanical environment leading to adaptation of tissues is also necessary in non-specialised cells to allow the organism to maintain the correct functional balance of its tissues. This is especially so in skeletal and heart muscle, blood vessels, bone, skin and lung epithelium. Having too much tissue in a high functional state is wasteful. Too little tissue which has low performance means that the organism cannot meet the demands of

Society for Experimental Biology Seminar Series 54: *Biomechanics and Cells*, ed. F. Lyall & A. J. El Haj. © Cambridge University Press 1994, pp. 197–219.

the environment. A sensing system with feedback thus operates in these tissues to maintain the optimal functioning tissue, which is balanced between breakdown and buildup. Whereas the mechanical loads on skin and muscle can result in strains of up to 150% on the cells, in bone the mechanical loading results in strains up to 0.5% (at which point it begins to fracture), but more typically 0.3%, as discussed elsewhere in this book. This places the strain transducing mechanism(s) of bone cells at amongst the most sensitive, nearly equalling that of specialised organs such as the hearing apparatus. The term 'strain' has been defined elsewhere in this book. In the first part of this chapter devices are described that are designed to apply homogeneous strains in the region of 100–500 000 microstrains (µstr) (0.01%–150%) to cells in culture. These strains can be tensile or compressive.

Many theories have been advanced as to the mechanisms by which cells can sense mechanical loading. In 1987 a mechanism, by which a membrane-bond phospholipase C was activated in osteoblasts by defined uniaxial strains, was first described (Jones & Scholübbers, 1987), which now appears to be a general mechanism of transduction of mechanical forces into biological response. We first describe methods on how to apply defined mechanical loading to cells. We speculate on what the mechanism of transduction in bone might be and on how information produced by mechanical loading might be passed on to other cells to achieve a coordinated tissue response.

Bone responds to physiological loading in several ways: increase in cortical thickness, increase in number and thickness of the trabeculae, an increase in bone mineral density (which probably reflects changes in the spaces around the osteocytes rather than an increase in calcium mineral density on the collagen matrix), remodelling of the internal structure (Haversian system and trabecular remodelling) and a strain-induced change in proteoglycans (Skerry et al., 1990). All of these result in adaptive changes so that the bone can withstand the actual stresses imposed on it. The principle of functional adaptation is often called Wolff's Law, although it is not a law in the scientific sense of the word, but rather a description of empirical findings. Mechanical loading of bone results in deformation. These deformations vary depending on the type of bone and the direction and amount of load placed upon it. Some people suggest that the piezoelectric and streaming potentials induced in bone during deformation result in physiological responses (Bassett & Becker, 1962; Gjeslvik, 1973; Lee et al., 1981; Davidovitch et al., 1984; Liboff et al., 1987) and some people suggest that hydrostatic forces can elicit cellular responses (Bagi & Burger, 1989; Hall, Urban & Gehl, 1991; Burger,

Klein-Nulend & Veldhuijzen, this volume), although at low pressures it is difficult to conceive of a physical mechanism to account for a biological response.

Using strain gauges and sophisticated models, many groups have described these deformations in a variety of vertebrates. As there is no way to sense load directly, only to infer it from the deformations produced in relation to the stiffness of the material, cells must either respond to the deformation (strain) or to the strain-related potentials (SRP). Cells cannot sense load, and must respond to deformations. This was first pointed out by Lanyon, who went on to determine the minimal effective strain to maintain bone mass (500 µstr) and the effective strain to increase it (1500 µstr) (Lanyon, 1984). Physiological responses were also found to peak strains of 3000 µstr which were produced at a strain rate of between 4000 and 18 000 µstr^{-1} (Rubin & Lanyon, 1984). Owing to the shape of bone the strains induced in the long bones are mainly uniaxial at any one time, but vary in direction over the loading cycle. The distribution of strain within the bone, as pointed out by Lanyon (this volume), is also an important controlling parameter of the response of bone to loading. It is very difficult to recreate exactly these strain conditions in culture, but the cell culture technique offers the possibility to control better the strain parameters and to investigate the transduction mechanisms in greater detail.

Previous investigations into the effect of strain on osteoblast cell cultures either have not applied physiological levels of strain, have not measured the strains applied, or have used apparatus that does not apply the same level of strain to al the cells in the culture. For instance, Harell, Dekel & Binderman (1977) investigated the production of prostaglandin E_2 (PGE$_2$) in cell cultures using a deformed plastic culture dish but did not measure the strains or distribution of strains applied to the cells. Yeh & Rodan (1984) used a cyclically stretched collagen ribbon to investigate PGE$_2$ synthesis and have applied 100 000 µstr at high strain rates. Hasagawa *et al.* (1985) applied a three-point bending system, that also does not apply equal strains to the culture surface, to investigate protein and DNA synthesis. Buckley *et al.* (1988) applied very high strains (up to 250 00 µstr) to osteoblasts and have reported a significant change in shape and a reorientation of the cells. Murray & Rushton (1990) used an apparatus that applied physiological levels of strain in a type of machine that produces equal strain over the culture surface, also to study PGE$_2$ synthesis in strained cells.

Machines for strain application

The first aim of our investigations was to subject all the osteoblast-like cells to the same physiological level of strain and strain rates. To do this we have designed and built several devices that meet these requirements. Using these machines we have conducted investigations into the likely mechanism of the strain transducer and the biological signal transduction cascade, from the second messenger system and electrophysiological phenomena to gap junctional communication and stimulation of growth factor mRNA synthesis. General requirements for all substrates used for strain experiments are that:

1   they are elastic in the region for which they are used;
2   cells attach to the surface; and
3   cells can grow and differentiate on the surface.

Other properties depend on the type of experiment, such as ease of extraction from the surface and verification of the extraction process, and whether special optical properties are required.

Four-point bending machine

The 4-point bending machine is based on the fact that between two fulcra a plate can be bent into a constant curve. Not only does geometrical analysis bear this out, but direct measurement of the surface by means of strain gauges also shows this. Since the strain gauge technology can be expensive to set up, a geometrical measurement that is as accurate can be used by laboratories without access to strain gauge amplifiers. We use plates made from toughened glass and from polycarbonate. We have developed the theory of 4-point bent plates to allow us to choose different bending moments from plates of different thickness so that a control for fluid movement effects (possible fluid shear effects and possible streaming potential effects) can be investigated and minimised. The calculation of the surface strain in 4-point bending is based on the geometry of the situation. Figure 1 shows a diagram of the principle of 4-point bending.

It is possible to calculate the strain on the surface from a simple formula, without the need for measuring the strain with a strain gauge. Since the plate is bent into a constant curve, a geometrical model can be used to describe the strains on the tension and compression surfaces. The tensile and compressive strains are, in most homogenous materials, equal. The middle axis of the plate has no strain and is called the neutral axis as shown in the figure. The strain on either surface depends on the distance between the supports (D), the thickness of the material (h) and the amount of deflection (y).

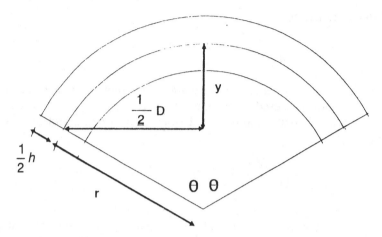

Fig. 1. Diagram of 4-point bending of culture plate. A plate is bent between two fulcra into an even arc of a circle. The dimensions are used in the equations to calculate strain, where: h = plate thickness = $2x$; D = distance between supports = $2a$; Y = amount of deflection; S = strain; r = radius. For further details see text.

Strain is defined as the change in length:

$$S = \frac{l' - l}{l}$$

From the figure we can derive the geometrical description as following:

$$l' = (r + x)\,\theta$$

$$l = r\theta$$

$$l' - l = x\theta$$

$$r^2 = a^2 + (r - y)^2 = a^2 + r^2 - 2ry + y^2$$

$$\theta = \frac{l}{r} = \frac{2yl}{a^2 + y^2}$$

$$l' - l = x\theta = \frac{2xyl}{\alpha^2 + y^2}$$

$$S = \frac{2xy}{a^2 + y^2}$$

which reduces to

$$S = \frac{4hy}{D^2 + 4y^2}$$

We have found that the calculation of strain given here correlates exactly with measured strain using strain gauges.

To find the deflection expected from a given strain and thickness of plate:

$$y = \frac{\frac{h}{S} - \sqrt{\left(\frac{h}{S}\right)^2 - D^2}}{2}$$

This last equation is useful to compare the amount of deflection required to produce a certain strain with different thicknesses of plates. The 4-point bending machine can therefore be used to produce the same strains under different degrees of bending. A thin plate needs to be bent more than a thick one to produce the same surface strain. This property allows the investigation of the effect of fluid flow induced by the bending, and possibly inducing streaming potentials by using a high bending and low strain against the same strain but low bending, as are described later. This feature is not possible with other types of strain devices, as are also described below.

The materials used for the plates are polycarbonate which has a relatively low Young's modulus compared to the 1 mm thick toughened glass plates used as described in an earlier publication (Jones *et al.*, 1991). The amount of force required to bend the plates is related to the Young's modulus and the dimensions of the plate. Polycarbonate can be strained up to 3000 µstr by using a lever mechanism attached to a 500 W electric motor in thicknesses up to 4 mm, as can 1.5 mm toughened glass plates (DESAG AG Grünenplan, Germany). The glass plates can be strained up to 4500 µstr at 3 Hz, but fracture when scratched. The 4-point bending system as described can produce strains of up to 10 000 µstr in 2.5 mm polycarbonate plates which appears to be the limit of the material and the machine. Medical grade culture silicone is used to make wells to contain the culture medium, attached to the surface using Dow Corning silicone aquarium sealer, well cured.

The *in vitro* 'biocompatibility' of the various materials was assessed by measuring the amount of non-collagenous matrix protein produced per cell. The more bone type matrix produced per cell, the more 'biocompatible' was the material judged to be. This quotient has been found a useful tool in evaluating the 'biocompatibilty' of biomaterials with a high degree

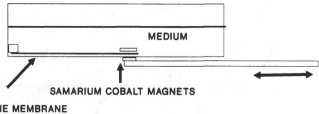

MEDIUM

SAMARIUM COBALT MAGNETS

SILICONE MEMBRANE
SURFACE MODIFIED

Fig. 2. Diagram of magnetically coupled hyperphysiological strain device. Strains of up to 150% (500 000 μstr) can be produced.

of correlation with animal evaluation (Jones & Scholübbers, 1987). At the same time the types of matrix proteins produced were identified to check for normal differentiation of the cells. It was found that the toughened glass plates cause a more rapid growth of the cells and a lower degree of differentiation within the 1 week of culture than either the normal glass or cell culture polystyrene. A number of treatments were used to increase the biocompatibility of the surfaces. For plastics, immersion in concentrated H_2SO_4 for several days was found to be a good treatment.

Silastic uniaxial strain devices

Normal medical grade silicone (Dow Corning Silastic MDX-4-4210 Medical Grade Elastomer), is non-toxic, but does not promote good attachment and development of osteoblasts on the surface. Other grades of silicone, used for instance in casting over electrical equipment, contain tin as a polymerising catalyst and are toxic. Medical grade silicone membranes 75 μm thick modified by attaching anionic groups to the surface were obtained from Dr R. Siegel, Bionic Surfaces, Würzburg, Germany (the exact method of producing these modifications is an industrial secret at present). These membranes were used in three different devices which produced strains of up to 10 000 μstr and from 5000 to 500 000 μstr. Figure 2 shows a diagram of one of these devices used to produce strains of up to 500 000 μstr.

The membrane is placed in a long narrow trough, which is just as wide as the membrane. A samarium cobalt magnet is glued to the membrane at one end using aquarium sealer. Underneath, on the friction surface a thin (0.2 m) disc of polycarbonate is also glued. A thin (0.2 mm) polycarbonate foil separates the culture system from the pulling magnet. The pulling magnet can be operated by hand, or attached to a cam driven by

204   D. B. JONES *et al.*

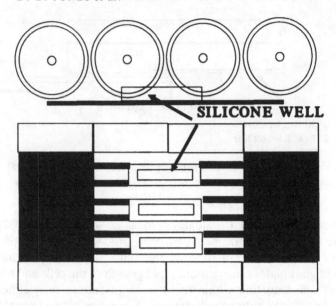

Fig. 3. Diagram of 'film spool' strain machine for large strains. Strains of up to 150% (500 000 μstr) can be produced in a small device.

a motor. Very good cell growth, attachment and development occurs on the modified silicone.

A second high strain device was also constructed based on what we call the film spool mechanism, suggested first by one of us (D.B.). In this device the thin silicone membranes are attached to the bottom of a silicone well using aquarium sealer. The silicone well is constructed of 1 mm thick medical grade silicone. As shown in Fig. 3, a flexible but non-elastic woven material is attached to the strain wells and to the underneath of the spools. The spools are connected to each other by cogs of which one pair is driven (either by hand or from a motor). The idler cog wheel translates the rotation so that the spools counterrotate symmetrically. A simpler version contains only one rotating spool.

The membranes and the culture containers are made of thin material to reduce the amount of energy required to stretch them; thus less power-ful and cheaper motors can be used for the actuation. These machines, as well as those described below, can also be used in compression if cells are seeded onto prestrained membranes. Relaxation of the membrane results in compressive strains on the cells. It is simpler to cycle between compression and strain in such a device than in the four-point bending system described above since the driving mechanism does not need to be very complicated.

Imaging during strain

The machines described above are useful for gathering data on large populations of cells. The area of the culture surfaces was between 1 cm^{-2} (for cell counting and protein/DNA measurements) and 10 cm^{-2} (for investigating second messengers, extraction of mRNA and measurement of collagen synthesis). It is possible to make many analyses using imaging techniques: not only DNA content, cell cycle analysis and cell counting, but also dynamic physiological measurements such as intracellular calcium concentration, membrane phospholipases, membrane potential, intracellular pH and also gene activation (using transfected cells containing luminescent and fluorescent reporter genes). To enable cells to be imaged during strain requires a different type of mechanism and also a suitable substrate on which the cells are strained. We have built two devices that meet these requirements. In addition to the specifications defined above (defined, measureable uniaxial strain on an elastic membrane) several further properties are needed.

1   The cell under observation must not move in space during strain, or the imaging system must be capable of following the cell during straining.

2   For certain types of reporter dyes the substrate must be non-fluorescent at the wavelengths used. If an inverse microscope is used, imaging must be possible through the thickness of the membrane.

3   For fluorescent inverse microscopes the substrate must be thin enough to allow the use of high magnification, high numerical aperture objectives.

We have used two types of machine to achieve these objectives and will describe them below.

Aclar cam stretching device

This device, shown in Fig. 4, produces strains of up to 5000 μstr. The actuating mechanism is a brass rod of square cross-section, machined so that the corners are rounded. The two cams are connected by a yoke that can be actuated by hand or by a motor. Owing to the shape of the cams the strains produced per angle of deflection are not linear, but can be calibrated using an optical measurement of displacement at a defined distance from the centre. The material ACLAR 33C (produced by Allied Signal, New Jersey, USA as a specialised wrapping material) has been used since the early 1980s for cell culture, especially as a substrate that

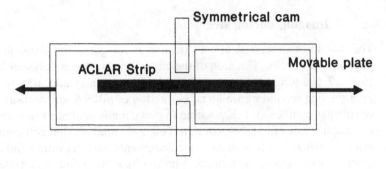

Fig. 4. Diagram of Aclar membrane cam stretching device. Symmetrical
stretching is obtained through mechanically coupled plates pushed apart
by a symmetrically shaped cam. Further details are in the text.

allows sectioning for electron microscopy. It is a terpolymer of fluoro-
carbons related to Teflon, is elastic up to 5000 μstr and transmits UV
light down to 300 nm without a great deal of disturbing fluorescence
(about $\frac{1}{1000}$ that of the FURA signal and less than autofluorescence of the
cells). Cells attach to the surface spread and also differentiate, especially
if the surface was pre-treated for 48 h with concentrated sulphuric acid.
The degree of differentiation and growth is, however, lower than for
culture plastic.

Magnetostrictive actuator device

Several devices exist using piezoelectric extension or the magnetostrictive
phenomenon, which results in a device capable of extending about
110 μm in a very controlled manner (± 25 nm). These devices can be
incorporated into a machine for applying well controlled strains to various materials. One problem with piezoelectric actuators concerns the
associated high electric fields, which have to be shielded if electrophysi-
ology equipment is used. Electric and magnetic field effects, which are
not discussed here, have been suggested to have physiological effects on
cells. The high electric and magnetic fields associated with piezoelectric
elements could cause a problem if these effects are significant. Another
problem is the relatively high cost of the piezoelectric translator.

In contrast the magnetostrictive actuator provides comparable extensions with lower costs and lower associated magnetic and electric fields.
The magnetostrictive actuator produces 110 μm of elongation, which
over 10 mm of membrane (about the smallest easily constructed culture
surface) gives 1.1% strain, and in symmetrical mode 2.2% (two opposing
actuators). Other devices using mechanical levers and hydraulic heads
can achieve larger elongations of up to 1 mm and 200 000 μstr at the

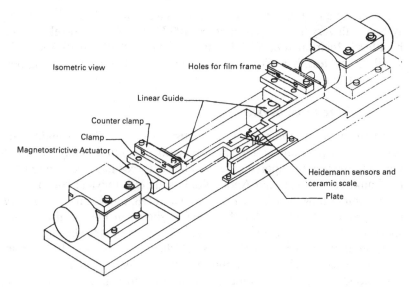

Fig. 5. Diagram of magnetostrictive actuator device. Symmetrical stretching is obtained through two magnetostrictive Terfenol actuators which pull the stretching frame apart. For further details see the text.

cost of frequency response and power. A magnetostrictive device using Terfenol elements (ETREMA Products Inc., Ames, Iowa, USA) has been designed by Dr Jos van der Sloten as part of a collaboration with the Biomechanics Department of Leuven. This device achieves much more precise deformations than the mechanical designs above. A diagram of the device is shown in Fig. 5. Shielding may be required in this device, as well as for the motors used in the other devices.

This device can stretch a 10 mm long silicone membrane up to 2.2% (22 000 μstr) at frequencies of up to 20 kHz. The cost of the control electronics for the two elements is about $4000. Included in the design is a laser interference positioning sensor, accurate to 100 nm, for feedback control of the actuators. The forces produced are very high (10 000 N), which means that the machine could also be used for other, stiffer, substrates.

At high strains the thinning of the membrane results in a change of focus as the cells drop towards the objective. The amount of out-of-focus movement depends on the strain applied and the thickness of the membrane. For a 50 μm thick silicone membrane strained to 120%, the material will thin by 10% (volume is preserved) which means that the upper surface of membrane will drop by 2.5 μm. This shift can theoretically be compensated for by linking the control electronics mechanism applying the strain to a motor or piezoelectric focusing system. A relatively cheap system with very high response times (100 Hz for 100 µm travel) is made by Physik Instrumente, Waldbronn, Germany (P-720.00 PIFOC). For small strains (less than 1%) this change in the focal plane does not appear to be significant.

Shear force devices

Fluid shear is physiological in some tissues, especially on endothelial cells, and this phenomenon might also be important in osteocytes. Rotating cone devices for applying defined shear forces have been described by Franke *et al.* (1984) and Ando, Komatsuda & Kamiya (1988). Parallel plate laminar flow devices by Frangos *et al.* (1985) and van Kooten *et al.* (1991), amongst others. Another device, shown in Fig. 6, also applies shear to cells by moving two surfaces against each other. This device was designed to investigate the effect of small shears occurring at the interface of two surfaces, such as are found in implanted biomaterials.

In blood vessels the cells modify themselves to maintain a fluid shear stress of 16 dyn cm^{-2} (Fung, personal communication), which is a parallel to 'Wolff's Law'. It is to be expected that fluid shear also results in cell deformations, but the SRP should be much higher. It would be interesting to compare the effects of applied shear and applied strain to elucidate the effects of SRP. One possibility, apart from the use of different thicknesses of plate in the four-point bending system as described above, would be to use ion-free osmotically balanced medium to increase the SRP by increasing the streaming potentials in both types of system, or to use an osmotically balanced medium containing fixed positive charges on a macromolecule, to decrease the streaming potentials. Another test of this theory would be to increase the viscosity of the fluid, so that a physiological shear would be obtained at lower flow rates, but the stream-

Shear Strain Applicator

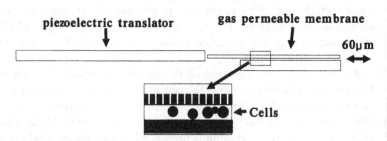

Fig. 6. Diagram of shear device for investigating the shear produced between two surfaces moving parallel to each other.

ing potential would be reduced. MacGinitie *et al.* (1987) tried to reduce the SRP by removing the hyaluronic acid residues, containing many fixed negative charges, from cartilage. However, this also affects the cells through the hyaluronic acid receptor and might account for the results obtained (Muir, 1977). By changing the streaming potential or the amount of fluid movement no difference was seen in the dynamics of intracellular calcium increase, which is thought to be linked to the first biochemical change associated with the strain sensor (see below). Increasing the viscosity of the fluid, without changing the ionic content, would also help to determine the source of the strain induced response, however this would also increase the streaming potential (ζ).

Other methods

Winston *et al.* (1989) have described a biaxial strain device which is based on a hydrostatic pressure (gas or liquid) induced extension of a disc of elastomer. The disc is subjected to hydrostatic pressure from underneath and thus expands in the shape of a dome. This device gives equal biaxial strain over the whole surface, is casy to use, is compact and can be configured for several simultaneous replicates. Brighton *et al.* (1991) have used this device for stretching osteoblast-like cells and in contrast to our previously reported results have found stimulation at 300 μstr. A simple model, which can perhaps account for this, is discussed below ('The Nature of the Transducer', p. 211).

Hereus petriperm plates may be stretched using this machine, but the elastic limit of the plastic used is below 1000 μstr. Murray & Rushton (1990) stretched Labtek plastic culture dishes uniaxially, which can withstand higher strains and maintains the silicone seal at 2000 μstr. Other mechanisms, such as those proposed by Hasagawa *et al.* (1985), Yeh & Rodan (1984), Harell *et al.* (1977) and Buckley *et al.* (1988), either do not reproduce the physiological strain environment or apply different strains to different parts of the substrate. Some devices are now available commercially that can apply loads to organ cultures or to cell cultures. El Haj & Thomas (this volume) have used compression of cells attached to cytodex beads to deform the cells, and have used the concept 'bulk strain' to describe the deformations produced (Shelton & El Haj, 1992).

Analysis of the strain sensor

What is the nature of the strain sensor in the osteoblast? Is there more than one sensor? What biochemical pathways are induced by strain? These questions are at present unanswered. Many authors now agree that one of the quickest responses to physiological levels of strain, albeit of different amplitudes for different types of cells, is an increase in intracellular free calcium. Jones & Scholübbers (1987, 1988, 1989), Jones *et al.* (1991) and Jones & Bingmann (1991) described the strain-related activation of phospholipase C (PLC) and rapid increase in intracellular free calcium (IFC) within 180 milliseconds of applied 3000 μstr. von Harsdorf *et al.* (1989) demonstrated stretch-activated PLC in heart muscle cells. Ando *et al.* (1988) observed a rapid increase in IFC in vascular endothelial cells when subjected to fluid shear stress both *in vitro* and *in vivo*. Wirtz & Dobbs (1990) showed IFC increases 12 s after applied stretches of 120% of biaxial strain in lung epithelial cells. The IFC increase described by Jones *et al.* (1991) was related to the PLC activation, but Jones & Bingmann (1991) provided evidence that shortly after the activation of a calcium-activated potassium channel, leading to hyperpolarisation of the cell membrane potential, a calcium channel was opened perhaps resulting from the increased IFC. After strain relaxation, the MP was again hyperpolarised, which was interpreted as a consequence of closure of the calcium channel and further metabolism of intracellular IP_3 until it was cleared from the cytoplasm. These results, gained with electrodes, were confirmed using membrane potential sensitive dyes and ratio imaging techniques, similar to that used for IFC imaging. Protein kinase C activation, which should also be expected due to PLC stimulation, was also found, confirming that PLC is located at or near the site of the strain transducing mechanism.

Owing to the fast response of IFC to strain, within the time resolution of the equipment used of 160 ms, it can be deduced that PLC is activated by strain before any other process. Thus the strain-related increase in PG synthesis reported by various groups (Palmer *et al.* (1973) in lung tissue, by the Bindermann group (Harell *et al.*, 1977; Somjen *et al.*, 1980; Bindermann *et al.*, 1988) in bone cells, by Murray & Rushton (1990), also in bone cells, and Sadoshima & Izumo (1993) in heart muscle cells) can be related to the prior activation of PLC and PKC directly. Whether other second messenger systems are activated directly by the strain transducer or are dependent on PLC activation is not yet known. Sadoshima & Izumo (1993) measured a stretch activation of phospholipase D, but 30 min after applying strain. The stimulation of other second messengers, such as tyrosine kinases, phospholipase D and $p21^{Ras}$ mitogen activated protein, were parameters also not resolved in time. The chain of events therefore remains to be established to deduce the existence of either one or more simultaneous pathways. If simultaneous pathways exist, then the question is whether they are linked to the same transducing mechanism. Stretch induced promotion of c-*fos* message reported by Komuro *et al.* (1991) is most likely linked to PLC–PKC activation (Sadoshima & Izumo, 1993). Hence it is not yet clear whether the transduction mechanism involves one or more mechanisms and one or more pathways. Since

Guharay & Sachs (1984) and Lansman, Hallam & Rink (1987) described stretch-activated ion channels, many groups have also investigated this phenomenon and described not only potassium channels, but also calcium channels. Duncan & Misler (1989) and Davidson, Tatakis & Auerbach (1990) have investigated mechanosensitive ion channels in osteoblasts. Stretch-activated ion channels have also been suggested to play a role in muscle response to stretch (Kirber, Walsh & Singer, 1988). As Morris & Horn (1991) did not find evidence to support the existence of such channels *in vivo* and taking into account the quick release of IFC, the evidence for a direct effect of strain on channel activation perhaps needs reviewing. An indirect effect on potassium channels and calcium channels through IFC release is supported by our data.

The nature of the transducer

The actual sensor of mechanical load appears from many pieces of evidence to be linked to the cytoskeleton. The first clue is that relaxation of many types of cell from their attachment site causes increases in IFC (reviewed by Ingber, 1991). Thurm (1983) has described changes in cytoskeleton structure during loading of the cockroach vibration sensor, where elastic proteins attached to the membrane and the cytoskeleton are stretched during loading. Since the cytoskeleton in many cells is under tension ('Tensegrity' of the cell, reviewed by Watson, 1991) of about 100 µdyn, it would seem logical that as for other structures, the most elastic parts of a cell will stretch the most when under load. These 'Thurm' structures, if they exist in other cells, would have the function of amplifying the stretch applied over the cell and localising this into a small area. Thus a stretch of 3000 µstr over a cell in culture of 40 µm in diameter would be a displacement in total of 120 nm. If the cell were mechanically homogenous, no significant stretch of any component would occur, but a Thurm structure would deform by about 30 nm. This deformation would be enough to dislocate a PLC inhibitor, attached to the cytoskeleton, physically from the PLC enzyme attached to the membrane and cause stimulation. Such a protein has been described in MCC (Mutated in Colorectal Carcinoma) a proto-oncogene (Kinzler *et al.*, 1991) and has been suggested as a possible candidate for a mechanotransducer (Jones & Bingmann, 1991). Thus, sites located closely to the attachment site, yet located at the cytoplasmic side of the plasma membrane, appear to be good candidates for such putative mechanisms. Such sites might be better candidates than, for example, the integrins, as suggested by Ingber (1991). Further evidence for the cytoskeleton associated mechanism comes from two further experiments. In one the IFC does not decrease under constant strain until after 10 min, which is about the time the filamentous actin cytoskeleton appears to depolymerise. This links the stretch-activated IFC to the integrity of the cytoskeleton. A further piece of evidence is that the aspect ratio of the cell (ratio of the diameter to the height) appears to control the sensitivity of the cell to compressive strains (Jones *et al.*, unpublished data), as described below.

Cell aspect ratio

According to the Thurm structure hypothesis, there will be a minimum effective strain which can activate the cell, as has been described in *in vivo* experiments. Analysis of the aspect ratio of the cells and the effect of compressive strain might be one way to determine part of the geometry of the strain sensor.

The aspect ratio of the cell is the ratio between the mean height and the mean diameter (or length on the strain direction). A cell will be expected to maintain its volume during a short strain (1–5 s). The preservation in volume means that an elongation in one or two axes will result in a reduction in length in the other dimension(s). Conversely, a compression in one or more axes will result in an elongation in another dimension, similar to squeezing a balloon. Uniaxial stretching of a cell 40 μm in diameter and 1 μm thick (a typical figure in cell culture) by 3000 μstr will result in an elongation of 120 nm and a decrease in height of about 1–3 nm (this figure depends on the shape factor of the cell, which is assumed here to be a disc). A compressive strain of 3000 μstr will cause a reduction in length of 120 nm but an increase in height of 1–3 nm. However, a cell that was 10 μm in diameter and 10 μm in height (more typical for *in vivo*) would increase by 30 nm. The hypothesis is that the strain is concentrated in the most elastic part of the cytoskeleton and that this distension is the trigger for the strain response (for example by removing an inhibitor of PLC). An increase of 3 nm might not be enough to trigger the response, and thus the aspect ratio of the cell will determine the cell's sensitivity to compressive strains. In one set of experiments we have shown that while tensile strains of 3000 μstr 30 cycles a day at 1 Hz caused an increase in cell number of 30% over 4 days, compressive strains of the same dimensions stimulated cell division by only 5%. This result also agrees with the prediction of 'Thurm Structures' within the osteoblast. Tuncay *et al.* (1989), working with chondrocytes in agarose, where the cells are round, have found no difference in cellular response to the same amplitude of tensile or compressive strain. We have attempted to change the aspect ratio of the cell in culture by using collagen surfaces where the cells attach and have a more rounded morphology. However, the collagen would not remain fixed to the surface when subjected to strain. Other methods are being investigated at present.

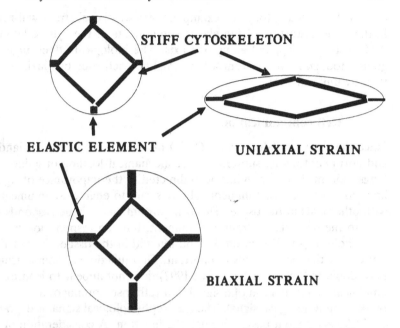

Fig. 7. Hypothetical scheme of two dimensional cytoskeleton deformations during single axis and biaxial deformations. Strain is additive through the cytoskeleton structure by biaxial strain in this model.

Biaxial stretching

Brighton *et al.* (1991) have applied biaxial loading to osteoblasts in culture and have detected physiological responses at strains of 300 μstr, which are lower than those thought to be stimulatory *in vivo*. Current concepts of the involvement of the cytoskeleton in the strain-sensing mechanism might predict this phenomenon. Figure 7 shows in diagrammatic form the possible explanation. We postulate that the cytoskeleton structures associated with the strain sensors have properties like springs and are under tension. At the attachment sites there are hinges so that the angle between cytoskeleton structures can change. Hence, if the cell is stretched along one axis it gets thin and long, the cytoskeleton will change shape and slightly relax the deformation in the postulated 'Thurm' elements. The exact amount of stretch will depend on the relative elasticity of the different parts of the structure and the stiffness of the hinge. A completely stiff cytoskeleton without such a structure would not display such an additive interaction. The mechanical properties of the cytoskeleton need further investigation to resolve this point. Thus biaxial stretching could result in an additive strain in the elastic components of the cytoskeleton. For example, 300 µstr biaxial strain will result in the same distortion of the elastic components of the cytoskeleton as if 600 µstr were applied along one axis. The biological effect might be greater too, since more strain sensors might be activated by biaxial strain than by uniaxial strain.

Cell communication

Since the work of Palmer *et al.* (1973) on the release of prostaglandins and aorta contracting substance after mechanical loading of guinea pig lungs, not much attention has been directed to the importance of signalling molecules released immediately by strain to enable communication with other cells in the tissue. This is important since tissue responds not only to the immediate strain environment, but can respond locally and as a whole organ. The strain distribution within the tissue, discussed by Lanyon in this book, is also important. Gap junctional communication in osteoblasts (Schirrmacher *et al.*, 1992) does not appear to lead to any amplification mechanism of a signal, but rather smoothing of a hormonal or electrophysiological signal. The electrophysiological signal is degraded by 50% each time it passes through the junction. A consideration of the rate of diffusion, reduced by the viscosity of the cytoplasm and the small pores of the gap junctions, would also suggest that significant signals are perhaps not passed amongst cells this way. However, the gap junctional network seems ideally suited for coordinating information about the distribution of strain within the tissue.

The work of Palmer *et al.* (1973) mentioned above, and that of Bodin *et al.* (this volume), suggest that the immediate release of signalling molecules might also play a significant role for short-term (minutes) modulation of the physiological response. Short-term release of signalling molecules might be mediated by the depolarisation of the membrane during calcium channel opening in a manner analogous to the mast cell degranulation. ATP is one of the many candidates since osteoblasts possess ATP receptors linked to PLC, thus causing a positive feedback loop, a prerequisite for signal amplification. Factors other than PGE_2 might also be released from osteoblast membranes, which is a point for further investigation. Although increase in TGFβ mRNA might be expected from a consideration of the signal transduction mechanisms stimulated by strain, we have not yet succeeded in detecting any changes. Other growth factors such as insulin-like growth factor II (IGF-II) could also be expected to be regulated by the strain-induced second messenger systems so far described. This process is dependent on mRNA synthesis, induced by the second messengers linked to the strain sensor.

Strain induced differentiation and de-differentiation

The strain environment appears to be able to change phenotypic expression of many types of cells. Applying hyperphysiological strains (to osteoblasts) of 101% and 130%, 30 cycles a day for 5 days causes a change in morphology and an orientation of the cells at 90° to the direction of strain (mentioned by Buckley *et al.*, 1988 using 125% strain). The higher strain level also caused the osteoblasts to lose osteocalcin expression, to increase type III collagen synthesis and to increase glycosaminoglycan synthesis by 700%. Yen, Yue & Suga (1989) have also noticed an increase in type III collagen synthesis *in vivo* in tissues subjected to hyperphysiological strains. Jones (1993) has suggested that the hyperphysiological strains found at shear interfaces between implants and tissue is one of the major causes of the fibrous capsule. Synovial cells have been reported to differentiate de novo at this interface (Revell & Lalor, 1992), perhaps induced by the strain environment.

Conclusions

Strains appropriate to the physiological strain environment of the cells can be reproduced in culture. Different machines that apply homogenous strains of different frequencies and amplitudes may be constructed to investigate different aspects of the strain response. Analysis of the sequence of events set in motion after strain can help to deduce the nature of the strain-sensing mechanism and the biological consequences. The first event, occurring at least 160 ms after the application of strain appears to be the activation of PLC, which might set in motion a sequence of biological mechanisms which coordinates many of the physiological responses to the mechanical environment. Other mechanisms, such as the release of immediate signalling molecules, the induction of transcription of growth factor mRNA and their translation into protein are long-term responses. Gap junctional communication might coordinate the response over the tissue as a whole. The strain sensor appears to be located in the cytoskeleton. No evidence for a significant role of streaming potentials has been discovered. Strains higher than those normally experienced in the body might evoke a specific change in phenotype and induce a type of pathological response.

Acknowledgements

This work was supported by the International BIOSIS society and by the Bertelsmann foundation. We are grateful to Prof P. Brinkmann, Institut für Biomechanik, University of Münster, for help in understanding physical concepts and to Herrn W. Rück in the Biomechanik workshop for translating many of the ideas for equipment into reality.

References

Ando, J., Komatsuda, T. & Kamiya, A. (1988). Cytoplasmic calcium response to fluid shear stress in cultured vascular endothelial cells. *In Vitro* **24**, 871–7.

Bagi, C. & Burger, E. H. (1989). Mechanical stimulation by intermittent compression stimulates sulfate incorporation and matrix mineralization in fetal mouse long-bone rudiments under serum-free conditions. *Calcified Tissue International* **45**, 342–7.

Bassett, C. A. L. & Becker, R. O. (1962). Generation of electric potentials by bone in response to mechanical stress. *Science* **137**, 1063–4.

Binderman, I., Zor, U., Kaye, A. M., Shimshoni, Z., Harell, A. & Somjen, D. (1988). The transduction of mechanical force into biochemical events in bone cells may involve activation of phospholipase A2. *Calcified Tissue International* **42**, 261–6.

Brighton, C. T., Strafford, B., Gross, S. B., Leatherwood, D. F., Williams, J. L. & Pollack, S. R. (1991). The proliferative and synthetic response of isolated calvarial bone cells of rats to cyclic biaxial mechanical strain. *Journal of Bone and Joint Surgery* **73A**, 328–31.

Brighton, C. T., Strafford, B., Gross, S. B., Leatherwood, D. F., Williams, J. L. & Pollack, S. R. (1991). The proliferative and synthetic response of isolated calvarial bone cells of rats to cyclic biaxial mechanical strain. *Journal of Bone and Joint Surgery* **73A**, 320–31.

Buckley, M. J., Banes, A. J., Levin, L. G., Sumpio, B. E., Sato, M., Jordan, R., Gilbert, J., Link, G. W. & Tran Son Tay, R. (1988). Osteoblasts increase their rate of division and align response to cyclic, mechanical tension *in vitro*. *Bone and Mineral* **4**, 225–36.

Canfield, J. & Eaton, C. (1990). Swimbladder acoustic pressure transduction initiates Mauthner-mediated escape. *Nature* **347**, 760–2.

Davidovitch, Y., Shanfeld, J. L., Montgomery, P. C., Lally, E., Laster, L., Furst, L. & Korostoff, E. (1984). Biochemical mediators of the effects of mechanical forces and electric currents on mineralised tissues. *Calcified Tissue International* **36**, 86–97.

Davidson, R. M., Tatakis, D. W. & Auerbach, A. L. (1990). Multiple forms of mechanosensitive ion channels in osteoblast-like cells. *Pflügers Archiv* **416** 646–51.

Duncan, R. & Misler, S. (1989). Voltage-activated and stretch-activated Ba^{++} conducting channels in an osteoblast-like cell line (UMR 106). *FEBS Letters* **251**, 17–21.

Frangos, J. A., Eskin, S. G., McIntyre, L. V. & Ives, C. L. (1985). Flow effects on prostacyclin production by cultured human endothelial cells. *Science* **227**, 1477–9.

Franke, R.-P., Gräfe, M., Schnittler, H., Seiffge, D., Mittermayer,

C. & Drenckhahn, D. (1984). Induction of human vascular endothelial stress fibres by fluid shear stress. *Nature* 307, 648–9.

Gehring, C. A., Williams, D. A., Cody, S. H. & Parish, R. W. (1990). Phototropism and geotropism in maize coleoptiles are spatially correlated with increases in cytosolic free calcium. *Nature* 345, 528–30.

Gjelsvik, A. (1973). Bone remodeling and piezoelectricity – I. *Journal of Biomechanics* 6, 69–77.

Golz, R. & Thurm, U. (1991). Cytoskeleton–membrane interactions in the cnidocil complex of hydrozoan nematocytes. *Cell and Tissue Research* 263, 573–83.

Guharay, F. & Sachs, F. (1984). Stretch activated K$^+$ channels in muscle cell membranes as mechanotransducers? *Journal of Physiology* 352, 685–701.

Hall, A. C., Urban, J. P. G. & Gehl, K. A. (1991). The effects of hydrostatic pressure on matrix synthesis in articular cartilage. *Journal of Orthopaedic Research* 9, 1–10.

Harell, A., Dekel, S. & Binderman, I. (1977). Biochemical effect of mechanical stress on cultured bone cells. *Calcified Tissue International* 22, 202–6.

von Harsdorf, R., Lang, R. E., Fullerton, M. & Woodcock, E. A. (1989). Myocardial stretch stimulates phosphotidylinositol turnover. *Circulation Research* 65, 494–501.

Hasagawa, S., Sato, S., Saito, S., Suzuki, Y. & Brunette, D. M. (1985). Mechanical stretching increases the number of cultured bone cells synthesizing DNA and alters their pattern of protein synthesis. *Calcified Tissue International* 37, 431–6.

Ingber, D. (1991). Integrins as mechanochemical transducers. *Current Opinion in Cell Biology* 3, 841–8.

Jones, D. B. (1993). The biomechanical origin of foreign body response. *Cells and Methods* 4 (in press).

Jones, D. B. & Bingmann, D. (1991). How do osteoblasts respond to mechanical stimulation? *Cells and Methods* 1, 329–40.

Jones, D. B., Nolte, H., Scholübbers, J.-G., Turner E. & Veltel, D. (1991). Biochemical signal transduction of mechanical strain in osteoblast-like cells. *Biomaterials* 12, 101–10.

Jones, D. B. & Scholübbers, J.-G. (1987). Evidence that phospholipase C mediates the mechanical stress effect in bone. *Calcified Tissue International* 41 (Suppl.), 4.

Jones, D. B. & Scholübbers, J.-G. (1988). Mechanical stress transduction in OB-like cells. *Fortschritte der Osteologie in Diagnostik und Therapie (Advances of osteology in diagnostics and therapy)*, ed. F. H. W. Heuck & E. Keck, pp. 382–9. Berlin: Springer-Verlag.

Jones, D. B. & Scholübbers, J.-G. (1989). A role for PI-phospholipase C in the medium of mechanical stress in bone surface cells. *Calcified Tissue International* 44 (Suppl.), S-97, p. 6.

Jones, D. B., Scholübbers J.-G., Becker, M. & Matthiass, H. H. (1988).

An osteoblast-like cell culture biocompatibility test of bone compatible materials. *Zeitschrift für Zahnärtzliche Implantologie* 4, 290–4.

Kinzler, K. W., Nishisho, I., Nakamura, Y. *et al.* (1991). Mutations of chromosome 5q21 genes in FAP and colorectal cancer patients. *Science* 251, 1366–70.

Kirber, M. T., Walsh, J. V. & Singer, J. J. (1988). Stretch-activated ion channels in smooth muscle: a mechanism for the initiation of stretch-induced contraction. *Pflügers Archiv* 412, 339–45.

Komuro, I., Kaida, T., Shibazaki, Y., Kurobayashi, M., Katoh, Y., Hoh, E., Takaku, F. & Yasaki, Y. (1990). Stretching cardiac myocytes stimulates protooncogene expression. *Journal of Biological Chemistry* 265, 3595–8.

Lansman, J. B., Hallam, T. J. & Rink, T. J. (1987). Single stretch-activated ion channels in vascular endothelial cells as mechanotransducers. *Nature* 325, 811–13.

Lanyon, L. E. (1984). Functional strain as a determinant for bone remodelling. *Calcified Tissue International* 36, 556–61.

Lee, R. C., Frank, E. H., Grodzinsky, A. J. & Roylance, D. K. (1981). Oscillatory compressional behaviour of articular cartilage and its associated electromechanical properties. *Journal of Biomechanical Engineering* 103, 280–92.

Liboff, A. R., Rozek, R. J., Sherman, M. L., McLeod, B. R. & Smith, S. d. (1987). $^{45}Ca^{++}$ cyclotron resonance in human lymphocytes. *Journal of Bioelectricity* 6, 13–22.

MacGinitie, L. A., Grodzinsky, E. H., Frank, E. H. & Gluzband, Y. A. (1987). Frequency and amplitude dependence of electric field interactions. Electrokinetics and biosynthesis. In *Mechanistic Approaches to Interactions of Electric and Electromagnetic Fields with Living Systems*, ed. M. Blank & E. Findl, pp. 133–50. New York: Plenum Press.

Morris, C. E. & Horn, R. (1991). Failure to elicit neuronal macroscopic mechanosensitive currents anticipated by single-channel studies. *Science* 251, 1246–9.

Muir, H. (1977). Structure and function of proteoglycans of cartilage and cell–matrix interactions. *Cell and Tissue Interactions* 12, 87–99.

Murray, D. W. & Rushton, N. (1990). The effect of strain on bone cell Prostaglandin E2 release: a new experimental method. *Calcified Tissue International* 47, 35–9.

Nagel, G., Neugebauer, D., Schmidt, B. & Thurm, U. (1991). Structures transmitting stimulatory force to the sensory hairs of vestibular ampullae of fishes and frog. *Cell and Tissue Research* 265, 567–78.

Palmer, M. A., Piper, P. J. & Vane, J. R. (1973). Release of rabbit aorta contracting substance (RCS) and prostaglandin induced by chemical or mechanical stimulation of guinea-pig lungs. *British Journal of Pharmacology* 49, 226–42.

Revell, P. & Lalor, P. (1992). Synovial cells at the interface with retrieved implants. In *The Bone–Biomaterial Interface*, ed. J. E.

Davies, pp. 342–8. Toronto, Canada: University of Toronto Press.

Rubin, C. T. & Lanyon, L. E. (1984). Regulation of bone formation by applied dynamic loads. *Journal of Bone and Joint Surgery* **66A**, 397–402.

Sadoshima, J. I. & Izumo, C. (1990). Mechanical stretch rapidly activates multiple cell signal transduction pathways in cardiac myocytes: potential involvement of an autocrine/paracrine mechanism. *EMBO Journal* **12**, 1681–92.

Schirrmacher, K., Schmitz, I., Winterhager, E., Traub, O., Brümmer, F., Jones D. B. & Bingmann, D. (1992). Characterisation of gap junctions between osteoblast-like cells in culture. *Calcified Tissue International* **51**, 285–90.

Shelton, R. M. & El Haj, A. J. (1992). A novel microcarrier bead model to investigate bone cell responses to mechanical compression *in vitro*. *Journal of Bone and Mineral Research* **7**, S403–7.

Skerry, T. M., Suswillo, R., El Haj, A. J., Ali, N. N., Dodds, R. A. & Lanyon, L. E. (1990). Load-induced proteoglycan orientation in bone tissue *in vivo* and *in vitro*. *Calcified Tissue International* **46**, 318–26.

Somjen, D., Binderman, I., Burger, E. & Harell, A. (1980). Bone remodelling induced by physical stress is prostaglandin E2 mediated. *Biochemica et Biophysica Acta* **627**, 91–100.

Thurm, U. (1983). Mechano-electric transduction. In *Biophysics*, ed. W. Hoppe, W. Lohmann, H. Markl & H. Ziegler, pp. 666–71. Berlin: Springer-Verlag.

Tuncay, O. C., Neuman, R. G. & Shapiro, I. M. (1989). Chondrocyte response to compressed and tensed substratum. *Calcified Tissue International* **44** (Suppl.), S-100, p. 18.

van Kooten, T., Schakenraad, J., van der Moi, H. & Bucophor, H. (1991). Detachment of human fibroblasts from FEP–teflon surfaces. *Cells and Methods* **1**, 307–16.

von Harsdorf, R., Lang, R. E., Fullerton, M. & Woodcock, E. A. (1989). Myocardial stretch stimulates phosphotidylinositol turnover. *Circulation Research* **65**, 494–501.

Watson, F. A. (1991). Function follows form: generation of intracellular signals by cell deformation. *FASEB Journal* **5**, 2013–19.

Winston, F. K., Macarak, E. J., Gorfien, S. f. & Thibault, L. E. (1989). A system to reproduce and quantify the biomechanical environment of the cell. *Journal of Applied Physiology* **67**, 397–405.

Wirtz, H. R. W. & Dobbs, L. G. (1990). Calcium mobilisation and exocytosis after one mechanical stretch of lung epithelial cells. *Science* **298**, 1266–9.

Yeh, Ch. K. & Rodan, G. A. (1984). Tensile forces enhance prostaglandin E synthesis in osteoblastic cells grown on collagen ribbons. *Calcified Tissue International* **36**, 67–71.

Yen, E. H. K., Yue, C. S. & Suga, D. M. (1989). Effect of force level on synthesis of type III and type I collagen in mouse interparietal suture. *Journal of Dental Research* **68**, 1746–51.

I. BINDERMAN

Role of arachidonate in load transduction in bone cells

Introduction

It is widely accepted that bone is a highly adaptive tissue that modulates its architecture (mass and structure) in response to its mechanical environment. The adaptive process is regulated through homeostatic pathways as well as epigenetic regulatory processes (Turner, 1992). In bone, as in all connective tissues, the main constituents are the cells and the extracellular matrix. The latter is composed of collagen fibres and ground substance rich in glycosaminoglycans which undergo mineralisation. The surfaces of bone are lined with osteogenic cells and layers of precursors which are at interface with bone marrow on the endosteal surface and with fibrous connective tissue and muscle tissue at the periosteum. The calcified bone matrix surrounds the osteocytes, which have numerous long cell processes in contact with those of other osteocytes, or with processes from the lining cells on the endosteum or on the periosteum. This kind of morphology creates a very potent communication system between the matrix and the cells, as well as, between the lining cells on the surface and their neighbouring tissues (marrow and ligaments) and between the lining cells and osteocyte network. It is most probable that this unusual combination of mineralised collagenous fibres and the cellular system determine bone's unique mechanical properties and its remodelling capacity.

The general anatomical form of a bone is inherent in the skeletogenic tissue. Muscular and gravitational stresses determine the mass and distribution of bone tissue. A change in the vector of force will usually activate the regional and local remodelling pathway to reduce or even eliminate the vector. Since bone tissue is mostly mineralised, it absorbs many of the changes in force or its direction, without stimulating the cellular

Society for Experimental Biology Seminar Series 54: *Biomechanics and Cells*, ed. F. Lyall & A. J. El Haj. © Cambridge University Press 1994, pp. 220–227.

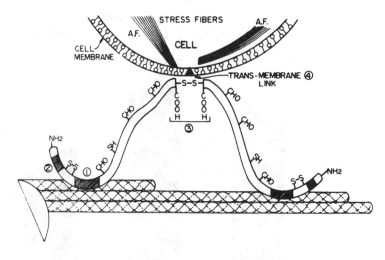

Fig. 1. Schematic presentation of chemical and mechanical attachment of osteoblasts to collagen bone matrix.

system. It is therefore the mechanical strain which activates the cell membrane and its enzyme population. This communication will describe the activation of enzymes and their environment by mechanical strain and the possible effects on different populations of cells in bone tissue. Remodelling of bone in response to mechanical perturbation is locally and regionally regulated most often by arachidonic acid metabolism, mainly Prostaglandin E2 (PGE_2).

Bone cells, osteoblasts and osteoclasts, are intimately related to the bone matrix. Osteoblasts produce the specific collagen fibres and control closely their direction and mineralisation. At the same time they attach to specific sites on the collagen molecule. The osteoclast is known to be attracted by mineralised bone matrix, attaching strongly to the matrix and forming a microenvironment for its resorbing activity. Mechanical strain of the matrix is able to produce a change in the cell membrane strain, activating some of the membrane enzymes (Fig. 1). In any cell membrane one might expect to find the various kinds of phospholipid molecules distributed in an orderly fashion. Two other lipids, glycolipids and cholesterol, are also found in the membranes of animal cells. The addition of cholesterol to the phospholipid matrix makes the membrane more rigid and less permeable. In fact, addition of cholesterol to osteoblasts in culture reduced their responsiveness to mechanical perturbation (Fig. 2). In addition, their response to parathyroid hormone (PTH) is diminished. It seems that membrane phospholipids are important in

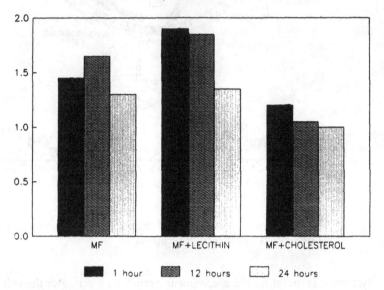

Fig. 2. The effect of cholesterol and lecithin on osteoblast responsiveness to mechanical stress. In this experiment osteoblasts in culture
were treated for 1, 12 and 24 h with cholesterol or lecithin and then
activated by mechanical perturbation as described in text. Cellular
cAMP was measured 20 min after mechanical perturbation.
Experimental/control levels of cAMP are expressed; 1.0 represents control levels and no effect of mechanical perturbation. This figure shows
that cholesterol-treated cells lost their responsiveness while lecithin
treatment is increasing responsiveness to mechanical perturbation.

transducing mechanical strain signals. It is interesting to note that the
cytoskeleton is also an important structure in transmitting the mechanical
signal. In order to study the mechanical signal at the biochemical cell
membrane level, a cell culture system of osteoblasts and their precursors
was developed (Binderman *et al.*, 1974).

Cell culture and method of application of mechanical stimulus

The experimental studies of Glucksmann over 55 years ago demonstrated
that static mechanical perturbation of bone and cartilage tissues *in vitro*,
devoid of blood supply, was able to mimic in principle the *in vivo* effects

of mechanical stress (Glucksmann, 1942). Rodan (Rodan *et al.*, 1975) was first to show that cells isolated from long bone epiphysis by enzymatic digestion revealed a cyclic AMP effect after being exposed to hydrostatic pressure. Our group (Binderman *et al.*, 1974) isolated cells from calvaria bone including periosteum by trypsin–EDTA digestion solution and cultured them for 3–4 weeks. The cells maintain their osteoblastic phenotype in culture during this period. Normally, the cells adhere to the surface of the culture dish and after a lag period of 24–48 h they proliferate, reaching confluency 5–7 days after plating. The seeding concentration is in the range of 500 000 cells per 60 mm cell culture Falcon dish. Like other normal fibroblast cells, the bone cells attach to the surface of the dish by pseudopodium-like extensions and make contact with neighbour cells via processes and membrane junctions. Recently, we were able to demonstrate dynamic cell communication – probably gap junctions – using fluorescent laser cytometry in these cultures (Binderman *et al.*, 1989). In our studies the mechanical perturbation is achieved by expanding an orthodontic expansion screw which is attached to two pieces of solid acrylic resin glued to the outer surface of the Falcon cell culture dish (Fig. 3). The expansion deforms the dish irreversibly (Harell, Dekel & Binderman, 1977). We were the first to show that synthesis of PGE_2 was an early step in the biochemical cascade of events in response to mechanical perturbation of confluent population of bone cells in culture (Somjen *et al.*, 1980).

Transduction mechanisms of the mechanical perturbation signal in bone cells

Recently, studies in our laboratory have suggested that mechanical strain of bone cells in culture activates phospholipase A2 which leads to release of arachidonic acid and the synthesis of prostaglandins (Binderman *et al.*, 1988). Indeed, addition of arachidonic acid (2 μg ml^{-1}) to the same cell cultures caused increased cyclic AMP production similar to that caused by mechanical perturbation of the dishes or addition of PGE_2. Also, when cultured bone cells were treated with phospholipase A2 (1 U ml^{-1} or 10 μg ml^{-1}) there was a two-fold increase in cAMP production, which was inhibited by antiphospholipid antibodies. Yet, inositol phospholipids also play an important role and their hydrolysis generates potent second messengers, like IP_3 which can mobilise calcium from intracellular stores and diacylglycerol, which can activate protein kinase C or be converted either to arachidonic acid or inositol phospholipid. Another inositol phospholipid, IP_4, is able to activate calcium channels in the cell membrane. It is inevitable that one or more of these messenger

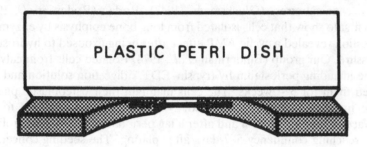

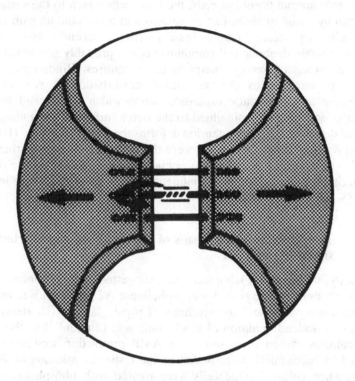

Fig. 3. Schematic presentation of the expansion device which is secured by epoxy glue to Falcon culture dishes. An orthodontic expansion screw is attached to two solid pieces of acrylic resin which forms the expansion device.

molecules activates the cell to synthesis and secretion of cytokines such as IGF-I, prostaglandins, and/or is involved in cell replication (Jones *et al.*, 1991; Sandy & Farndale, 1991). In addition, PGE_2, which is synthesised by osteoblasts in response to mechanical strain, is further secreted and locally and regionally may activate lining bone cells, mesenchymal cells and marrow cells. It was proposed by us and other investigators (Yeh & Rodan, 1984; Murray & Rushton, 1990; Rawlinson *et al.*, 1991) that PGE_2 is the main transduction mediator of mechanical strain in bone. PGE_2 then binds to a receptor on the cell surface and activates the remodelling cascade: effect on cAMP production (Somjen *et al.*, 1980), ornithine decarboxylase activity (Somjen *et al.*, 1982), creatine kinase activity (Somjen, Kaye & Binderman, 1985) and finally DNA synthesis. Many cell surface receptors exert their actions through specific G proteins; the involvement of G proteins in signal pathways to be activated by mechanical stress is also likely.sandy (Sandy & Farndale, 1991) suggested that the stimulatory Gs and inhibitory Gi, G proteins almost certainly control activation of adenylate cyclase in response to mechanically induced prostaglandin production. It is noteworthy that no receptor protein sensitive to mechanical perturbation, other than phospholipase A2, was described. However, it is possible that like the transducing receptor protein bound to G protein which is activated by photon energy in the eye and activates phospholipase A2 (Birnbaumer, 1990) a mechanoreceptor specific protein in the osteoblastic lineage exists. This is in agreement with published data which suggest that osteoblasts possess a G protein-sensitive phosphoinositide specific phospholipase C. G proteins modulate not only cAMP formation, but also intracellular ionic calcium mobilisation, arachidonic release, and membrane potential (Birnbaumer, 1990).

On the other hand, mechanical perturbation of skeletal cells from mandible condyle skeletal cells (Shimshoni *et al.*, 1984) and epiphyseal cartilage cells is activating the adenyl cyclase pathway not through synthesis of arachidonic acid and PGE_2. It seems likely that membrane changes associated with mechanical perturbation could be cell specific. More evidence for this hypothesis comes from studies using a non-invasive pulsed electric stimulation, coupled capacitatively to bone cells in culture (Binderman *et al.*, 1985). Using this technique, we were able to demonstrate immediate changes, following a short stimulation, in intracellular levels of cAMP changes, calcium uptake and the state of cellular actin. Moreover, a later increase in DNA synthesis has been observed under these conditions. All these changes were correlated with the strength of the applied electric field and the findings suggest that the response to various electric field intensities is cell specific.

226   I. BINDERMAN

Many *in vivo* studies showed that exogenous PGE_2 is activating bone modelling and remodelling in favour of bone formation, both in normal and osteopenic animals (Jee *et al.*, 1991; Mori, Jee & Li, 1992). Exogenous PGE_2 treatment of primary osteoprogenitor cell cultures (Yahav *et al.*, unpublished data) revealed an increase of the progenitor pool which progresses to osteoblastic phenotype.

References

Binderman, I., Duksin, D., Harell, A., Sachs, L. & Katchalski, E. (1974). Formation of bone tissue in culture from isolated bone cells. *Journal of Cell Biology* **61**, 427–39.

Binderman, I., Somjen, D., Shimshoni, Z., Levy, Y., Fischler, H. & Korenstein, R. (1985). Stimulation of skeletal derived cell cultures by different electric field intensities is cell specific. *Biochimica et Biophysica Acta* **844**, 273–9.

Binderman, I., Zor, U., Kaye, A. M., Shimshoni, Z., Harell, A. & Somjen, D. (1988). The transduction of mechanical force into biochemical events in bone cells involve activation of phospholipase A2. *Calcified Tissue International* **42**, 261–7.

Binderman, I., Aviv, V., Melamed, N. & Rahamimoff, R. (1989). The usage of fluorescence laser cytometry in the study of dynamic cell–cell communication in osteoblasts (Abstract). Jerusalem: Israel Society of Physiology.

Birnbaumer, L. (1990). G proteins in signal transduction. *Annual Review of Pharmacology and Toxicology* **30**, 675–705.

Glucksmann, A. (1942). The role of mechanical stresses in bone formation *in vitro*. *Journal of Anatomy* **76**, 231–9.

Harell, A., Dekel, S. & Binderman, I. (1977). Biochemical effect of mechanical stress on cultured bone cells. *Calcified Tissue Research* **22** (Suppl.), 202–9.

Jee, W. S. S., Mori, S., Li, X. J. & Chan, S. (1991). Prostaglandin E2 enhances cortical bone mass and activates intracortical bone remodeling in intact and ovariectomized female rats. *Bone* **11**, 253–66.

Jones, D. B., Nolte, H., Scholübbers, J. G., Turner, E. & Veltel, D. (1991). Biochemical signal transduction of mechanical strain in osteoblast-like cells. *Biomaterials* **12**(2), 101–10.

Mori, S., Jee, W. S. S. & Li, X. J. (1992). Production of new trabecular bone in osteopenic ovariectomized rats by prostaglandin E2. *Calcified Tissue International* **50**, 80–7.

Murray, D. W. & Rushton, N. (1990). The effect of strain on bone cell prostaglandin E2 release: a new experimental method. *Calcified Tissue International* **47**, 359–9.

Rawlinson, S. C. F., El Haj, A. J., Minter, S. L., Travers, I. A., Bennett, A. & Lanyon, L. E. (1991). Loading-related increases in prostaglandin production in cores of adult canine cancellous bone *in vitro*: a role for prostacyclin in adaptive bone remodeling? *Journal of Bone and Mineral Research* **6**, 1345–57.

Rodan, G. A., Bourret, L. A., Harvey, A. & Mensi, T. (1975). 3'5'Cyclic AMP and 3'5'Cyclic GMP: mediators of the mechanical effects on bone remodeling. *Science* **189**, 467–9.

Sandy, J. R. & Farndale, R. W. (1991). Second messengers: regulators of mechanically-induced tissue remodelling. *European Journal of Orthodontics* **13**, 271–8.

Shimshomi, Z., Binderman, I., Fine, N. & Somjen, D. (1984). Biochemical characterization of cells derived from the candyle head of young rats. *Archives Oral Biology* **29**, 827–31.

Somjen, D., Binderman, I., Berger, E. & Harell, A. (1980). Bone remodelling induced by physical stress is prostaglandin E2 mediated. *Biochimica et Biophysica Acta* **627**, 91–100.

Somjen, D., Yariv, M., Kaye, A. M., Korenstein, R., Fischler, H. & Binderman, I. (1982). Ornithine decarboxylase activity in cultured bone cells is activated by bone seeking hormones and physical stimulation. *Advances in Polyamine Research* **4**, 713–18.

Somjen, D., Kaye, A. M. & Binderman, I. (1985). Stimulation of creatine kinase BB activity by parathyroid hormone and by prostaglandin E2 in cultured bone cells. *Biochemical Journal* **225**, 591–6.

Turner, C. H. (1992). Editorial: Functional determinants of bone structure: beyond Wolff's Law of bone transformation. *Bone* **13**, 403–9.

Yeh, C. K. & Rodan, G. A. (1984). Tensile forces enhance prostaglandin E synthesis in osteoblastic cells grown on collagen ribbons. *Calcified Tissue International* **36**, S67–71.

N. PENDER and C. A. G. McCULLOCH

Effects of mechanical stretch on actin polymerisation in fibroblasts of the periodontium

Introduction

Periodontal tissues provide support and attachment for the teeth. The periodontium includes two mineralising connective tissues, alveolar bone and cementum, and two soft connective tissues, periodontal ligament and the lamina propria of the gingiva. The fibroblast is the predominant cell type in the soft periodontal connective tissues. This secretory cell synthesises extracellular matrix proteins including fibronectin, glycosaminoglycans (Hassell, Kimura & Hascall, 1986; Bartold, 1987) and a large array of collagens that are the most abundant structural proteins of periodontal connective tissues (reviewed by Narayanan & Page, 1983).

Remodelling of periodontal tissues is up-regulated by the application of mechanical forces which have been demonstrated to increase collagen turnover rates (Birkedal-Hansen, 1988; Sorsa et al., 1992) and to produce elevated levels of chondroitin sulphate in the fluid that drains periodontal tissues (Samuels, Pender & Last, 1993). Rapid turnover of collagens in the matrix of both gingiva (Page & Ammons, 1974) and periodontal ligament (Sodek, 1977) is also essential for continuous attachment of the roots to the alveolar bone.

To balance collagen synthesis and maintain the steady-state, collagen degradation must occur. Two pathways of collagen degradation have been identified: an extracellular collagenase-dependent route and an intracellular pathway independent of collagenase (Murphy & Reynolds, 1985). Notably, conditions of increased physical stress on the periodontal ligament as a result of the mechanical effects of orthodontic force increase collagenase activity in periodontal tissues (Sorsa et al., 1992).

Collagen turnover is mediated also by fibroblast phagocytosis (Ten Cate, Deporter & Freeman, 1976; Melcher & Chan, 1981). Currently, it

Society for Experimental Biology Seminar Series 54: Biomechanics and Cells, ed. F. Lyall & A. J. El Haj. © Cambridge University Press 1994, pp. 228–243.

is not known how remodelling of matrix proteins that is mediated by phagocytosis provides adaptation of the periodontium to applied physical forces.

As phagocytosis is an actin-dependent process, fibroblasts from periodontal tissues are good candidates to study how physical forces regulate actin assembly particularly the actin-rich, sub-membrane cortex (Stossel, 1989).

To study actin polymerisation in the early response of fibroblasts to mechanical deformation, cells were attached to flexible substrata and both F-actin and G-actin were quantitated. Morphometry was performed to assess cell shape. The dependency of actin assembly on GTP-binding proteins and calcium concentration was investigated.

Cell cultures and mechanical stretching

Fibroblasts derived from human gingiva and human periodontal ligament were obtained using the methods of Brunette, Melcher & Moe (1976) and were grown to subconfluence in T-75 flasks (Falcon, Becton Dickinson, Mississauga, Ontario) as described previously (Pender & McCulloch, 1991). Only cells between the third and tenth passage were used. Approximately 14 h before each experiment, cells were trypsinised and plated in single cell suspension at 10^5 cells in 60 mm dishes with a flexible, hydrophilic plastic growth surface (Petriperm, Bachofer GmbH, Reutlinger, Germany). This low plating density successfully reduced the likelihood of cell-to-cell contacts which would add additional complexity to the study of cytoskeletal dynamics after membrane deformation (Kolega, 1986).

The Petriperm dishes were mechanically stretched by extending them over convex plastic surfaces which conformed to the diameter of the Petriperm dish. Six interchangeable plastic convex shapes with various curvatures permitted stretching of the base of the Petriperm dishes by 0, 1, 1.2, 3.3, 6.8 and 9.2%, respectively. Stretching was carried out in a laminar flow hood under sterile conditions and cells were fixed with formaldehyde.

Determination of F-actin and morphometry

After fixation, cells were stained with 5×10^{-6} M FITC-phalloidin and the fluorescence from individual cells was measured with a microscope fluorimeter using a fixed area mask in the optical path. Fluorimetry measurements of FITC-phalloidin were directly proportional to the concentration of fluorescent species, and in measurements of FITC-phalloidin at the concentration used, were proportional to concentrations of cellular

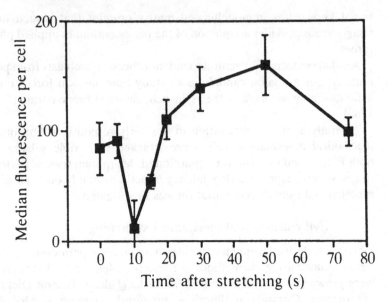

Fig. 1. Median fluorescence ± inter-quartile range (IQR) of human
gingival fibroblasts over a timecourse from 0 s up to 75 s when stretched.

F-actin (Pender & McCulloch, 1991). Gingival fibroblasts exhibited variations in F-actin content (Fig. 1) that were dependent on the time after
stretching. After 10 s of a 1.0% stretch, there was a three-fold reduction
in F-actin. After 60 s, there was a two-fold increase over baseline with a
return by 75 s to baseline levels of F-actin. When these determinations
were made contemporaneously using both the 1.0% and the 1.2% stretch
for up to 105 s, the dynamics of the response differed (Fig. 2). The initial
decrease in F-actin was greater and occurred more rapidly with the 1.2%
stretch. The increase in F-actin content at 50 s was greater. However, by
75 s, fluorescence levels returned to baseline.

    On the basis of these observations, two additional experiments (Fig.
3) were conducted in which cells were stretched 1.2% for different times
and allowed to relax. Relaxation for 10 s after a 60 s stretch led to a
two-fold increase in F-actin, relaxation for a further 20 s produced a
comparable increase in F-actin. Using a similar approach, relaxation for
10 s after a 90 s stretch produced a small increase in F-actin in excess of
the 90 s stretch value, relaxation for a further 20 s led to a fall in F-actin
levels close to the baseline value (Fig. 3). In a separate experiment (Fig.
3), stretch for 60 or 90 s followed by relaxation at three 10 s increments

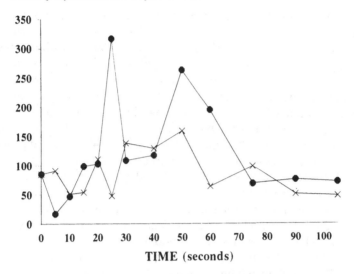

Fig. 2. Median fluorescence of human gingival fibroblasts over a time-course from 0 s up to 105 s when either stretched 1.0% (×) or 1.2% (•).

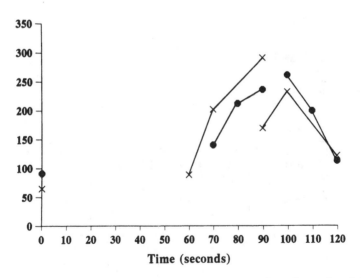

Fig. 3. Median fluorescence of human gingival fibroblasts after 1.2% stretch. × = stretch 60 s; relax 10 s, 30 s: stretch 90 s; relax 10 s, 30 s. • = relax 10 s, 20 s, 30 s after 60 s stretch; relax 10 s, 20 s, 30 s after 90 s stretch.

up to 30 additional seconds produced increases in F-actin up to an accumulated 90 s and a decrease from 100 to 120 s. Fluorescence returned to baseline values by 120 s. From these data relaxation after stretching seemed to induce actin assembly whilst conforming to the time-dependent variation of F-actin levels described in Figs. 1 and 2.

Notably, cultured osteoblasts subjected to 3 cycles min^{-1} of a maximum 24% elongation with a 10 s stretch followed by a 10 s relaxation produced a 50% increase in F-actin levels (Buckley *et al.*, 1990). Human skin fibroblasts attached to stretched collagen gels for 48 h showed prominent actin filament bundles which collapsed after 1 h of relaxation, the skin fibroblasts contracted and cell pseudopodia were retracted (Mochitate, Pawelek & Grinnel, 1991).

The dependence of actin assembly on the amount of stretch was invest-igated. Gingival fibroblasts (Fig. 4) exhibited the largest increase of actin assembly when stretched 3.3%. Periodontal ligament fibroblasts (Fig. 5) showed the greatest F-actin concentration at 0% stretch and the least at 6.8%. We found for the lines studied here that the fluorescence intensity per μm^2 due to FITC-phalloidin was three times greater for unstretched periodontal ligament fibroblasts than gingival fibroblasts, consistent with the findings of rich microfilament arrays in unstretched periodontal liga-

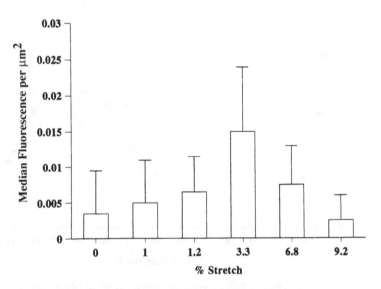

Fig. 4. Histogram of median fluorescence per μm^2 (+ IQR) of human gingival fibroblasts either unstretched (0) or stretched between 1% and 9.2% for 60 s.

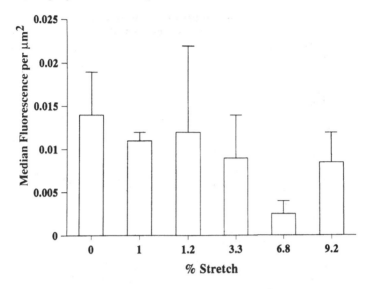

Fig. 5. Histogram of median fluorescence per μm^2 (+ IQR) of human periodontal ligament fibroblasts either unstretched (0) or stretched between 1% and 9.2% for 60 s.

ment fibroblasts (Beertsen, Everts & van den Hoof, 1974; Bellows, Melcher & Aubin, 1981). Unlike the increased assembly of actin in gingival fibroblasts subjected to stretch, periodontal ligament cells exhibited reduced F-actin after stretch.

Cell shape

The cell shape of attached human periodontal ligament fibroblasts differed from human gingival fibroblasts. In addition to the greater fluorescence intensity per μm^2 of unstretched periodontal ligament fibroblasts (Fig. 5) when compared with gingival fibroblasts (Fig. 4), image analysis of Toluidine Blue-stained cells showed the periodontal ligament cells to be longer (Fig. 6) with double the surface area (Fig. 7) of gingival fibroblasts. Stretching these cell types for 60 s showed a relationship between cell shape and the amount of stretch. Increased stretch tended to elongate periodontal ligament fibroblasts and shorten gingival fibroblasts. The periodontal ligament fibroblasts (Fig. 8) consistently showed a greater area/length ratio.

Examination of individual gingival fibroblasts by videocinemicrography

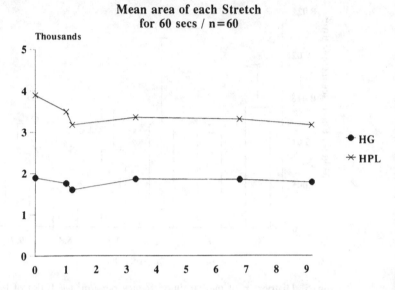

Fig. 6. The mean area (± sem) of human gingival (•) and human periodontal ligament (×) fibroblasts plotted against the percentage of stretch
applied for 60 s.

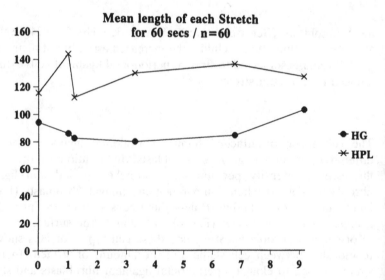

Fig. 7. The mean length (± sem) of human gingival (•) and human
periodontal ligament (×) fibroblasts plotted against the percentage of
stretch applied for 60 s.

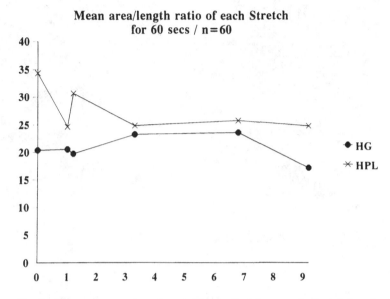

Fig. 8. The mean area/length ratio (± SEM) of human gingival (●) and human periodontal ligament (×) fibroblasts plotted against the percentage of stretch applied for 60 s.

(Fig. 9) over the first 60 s of stretch confirmed that at 1% stretch no detectable change in cell area or length occurred. However, at 180 s after stretching, large-scale changes in cell shape were evident including retraction of pseudopods and cell elongation. When comparisons of FITC-phalloidin fluorescence intensity per unit area were made between the leading and trailing portions of gingival fibroblasts (Fig. 10) with 1.2% stretch over a timecourse, there were no apparent differences within the cells. However, the time-dependent response to stretching in each subcellular site was similar as described above.

Total actin determination

Estimates of total actin were obtained by SDS-PAGE (Bray & Thomas, 1975) and densitometric quantitation of 41 kDa bands co-migrating with actin standards. Data from four dishes each of gingival and periodontal ligament fibroblasts at 10^5 cells per dish were analysed. Periodontal ligament fibroblasts that were not stretched contained 298 ± 97% more total actin than unstretched gingival fibroblasts, consistent with results obtained by microfluorimetry. Flow cytometric analyses of cell suspensions made after staining for both G-actin and F-actin (Knowles &

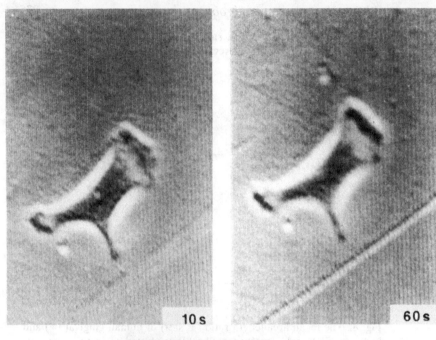

Fig. 9. Photomicrographs of videocinematography taken 10 s, 60 s and 180 s after 1.0% stretch of a human gingival fibroblast.

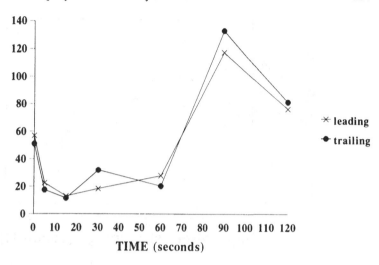

Fig. 10. Median fluorescence of human gingival fibroblasts over a time-course from 0 s up to 120 s stretched 1.2% and measured at the leading (×) and trailing (•) portions of the cell.

McCulloch, 1992) showed that periodontal ligament fibroblasts had a relatively larger F-actin content which was not simply due to a larger cell size.

G-actin determinations

Monomeric (G) actin was measured using modifications of the DNAase 1 inhibition assay (Blikstad *et al.*, 1978) described by Heacock & Bamburg (1983). Cell lysis was performed with 1% Triton in 15% glycerol, 10 mM Tris HCl, 2 mM MgCl$_2$, 0.2 mM DTT at -10 °C. Absorbance of DNA (50 mg ml^{-1}) was measured at 250 nm. The increase in absorbance of the solution was monitored at 10 s intervals during the first 60 s of the reaction and a best fit to the data points was made using a linear regression model (SAS, 6.03, SAS Institute, Cary NC). Data from two separate experiments using dishes plated at a cell density of 10^5 cells in 60 mm dishes were analysed. Rabbit muscle actin (Spudich & Watt, 1971) was prepared for standards. The variations in cellular F-actin as a function of time up to 60 s after stretching were associated with reciprocal variations in G-actin (Fig. 11), indicating the conversion of monomeric to filamentous actin (Korn, 1978) had occurred as expected. Consistent with previous reports (Howard & Meyer, 1984), the amount of increase of F-actin was not balanced by equally sized reductions of G-actin. This

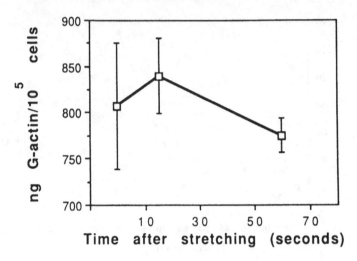

Fig. 11. G-actin content in ng after 1% stretching for up to 60 s.

probably represents the low sensitivity of the DNAase 1 inhibition to detect G-actin when cells are seeded at low plating densities.

Spatial relationships of actins

Gingival and periodontal ligament fibroblasts were examined by confocal microscopy and single cell fluorescence measurements were made of sub-strate-attached cells. The G-actin pool was stabilised and cells incubated with either DNAse 1 conjugated with tetramethylrhodamine isothio-cyanate (Rh-DNAse 1) or DNAse 1 conjugated with FITC (FITC-DNAse 1) incubated with FITC-phalloidin. Confocal microscopy demon-strated G-actin and F-actin were spatially separate in substrate-attached cells. Fluorescence intensities obtained by spectrophotometry (Table 1) demonstrated that unstretched periodontal ligament fibroblasts had higher levels of F-actin than unstretched gingival fibroblasts. After desta-bilisation of F-actin by azide, there was a reciprocal relationship between F-actin and G-actin and there was a relatively lower amount of G-actin in periodontal ligament fibroblasts than in gingival fibroblasts.

Regulation of actin assembly

Human gingival fibroblasts (Knowles & McCulloch, 1992) were incub-ated with either high concentrations of ATP (Heacock & Bamburg, 1983) or sodium azide (Hollenbeck et al., 1989). These methods of blocking

Table 1. *Single-cell fluorescence spectrophotometry of actin*

	Human gingival		Human periodontal ligament	
Actin type	F	G	F	G
Stain	FITC-phall	Rh-DNAse 1	FITC-phall	Rh-DNAse 1
Substrate-attached cells	10.1 (1.68)[b]	4.6 (0.43)	12.9 (1.34)[b]	3.8 (0.62)
Ratio FITC-phall to Rh-DNAse 1	2.2[a]		3.4[a]	

Mean photometer units ($\pm$ SD) ($n = 30$).
[a] = significantly different $p < 0.05$.
[b] = significantly different $p < 0.01$.
(From Knowles & McCulloch (1992).)

actin polyermisation and promoting depolymerisation both led to reduced amounts of F-actin and increased G-actin. When incubated with cytochalasin D for 1 h prior to stretching to block actin polymerisation, these cells, in common with periodontal ligament fibroblasts, showed no increase in the amount of F-actin after a 60 s stretch. Unstretched periodontal ligament fibroblasts, which had a greater baseline F-actin concentration than gingival fibroblasts, exhibited reduced levels of F-actin after incubation for 1 h with cytochalasin D.

To examine the dependence of actin assembly on guanine nucleotide binding proteins, human gingival fibroblasts were incubated for 2 h before stretching with pertussis toxin at 500 ng ml^{-1}. Stretching for 60 s showed no increase in F-actin. This blockage of stretch-activated actin assembly suggests a mechanism of signal transduction operating through guanidine nucleotide binding proteins (Taylor, 1990). Pretreatment of cells with either 10 mM EGTA or a medium containing no exogenous calcium before stretching both resulted in blockage of stretch-activated actin assembly.

Human gingival fibroblasts loaded with 2 μM fura2AM in buffer without serum and plated onto Petriperm dishes for 2 h before a 3.3% stretch exhibited a single peak of intracellular calcium 60 s after the stretch (Fig. 12). If this process was repeated with the cells incubated in a medium containing 0.5 μM EGTA (Fig. 13), there was no detectable calcium flux, indicating that the flux originated from extracellular calcium. Thus, calcium flux may be important in stretch-induced actin polymerisation similar to the involvement in cell shape changes observed in phagocytosis

240 N. PENDER & C. A. G. McCULLOCH

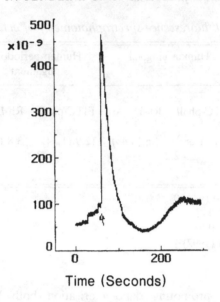

Fig. 12. Intracellular calcium measured after loading with fura2AM, human gingival fibroblasts stretched 3.3%.

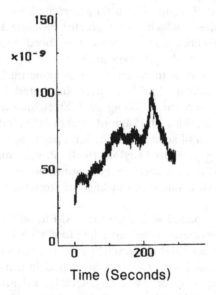

Fig. 13. Intracellular calcium measured after loading with fura2AM, human gingival fibroblasts incubated in medium with 0.5 μM EDTA and stretched 3.3%.

(Greenberg *et al.*, 1991). Notably, a variety of stimuli that activate exocytosis also induce actin reorganisation in chromaffin cells (Burgoyne, Morgan & O'Sullivan, 1989; Vitale *et al.*, 1991), mast cells (Koffer, Tatham & Gomperts, 1990) and parotid acinar cells (Perrin *et al.*, 1992).

Conclusions

The microfilament system of fibroblasts exhibits rapid dynamic responses to mechanical deformation prior to cell shape changes and appears to be dependent on cell type. Cell stretching stimulates actin assembly by a pathway of signal transduction operating through guanine nucleotide binding proteins and is dependent on an influx of extracellular calcium ions mediated by stretch-activation of calcium-permeable channels.

References

Bartold, P. M. (1987). Proteoglycans of the periodontium: structure, role and function. *Journal of Periodontal Research* **22**, 431–44.

Beertsen, W., Everts, V. & van den Hoof, A. (1974). Fine structure of fibroblasts in the periodontal ligament of the rat incisor and their possible role in tooth eruption. *Archives of Oral Biology* **19**, 1087–98.

Bellows, C., Melcher, A. H. & Aubin, J. E. (1981). Contraction and organisation of collagen cells by cells cultured from periodontal ligament, gingiva and bone suggest functional differences between cell types. *Journal of Cell Science* **50**, 299–314.

Birkedal-Hansen, H. (1988). From tadpole collagenase to a family of matrix metalloproteinases. *Journal of Oral Pathology* **17**, 445–51.

Blikstad, I., Markey, F., Carlsson, I., Persson, T. & Lindberg, V. (1978). Selective assay of monomeric and filamentous actin in cell extracts using inhibition of deoxyribonuclease 1. *Cell* **15**, 935–43.

Bray, D. & Thomas, C. (1975). The actin content of fibroblasts. *Biochemical Journal* **147**, 221–8.

Brunette, D. M., Melcher, A. H. & Moe, H. (1976). Culture and origin of epithelium-like and fibroblast-like cells from porcine periodontal ligament explants and cell suspensions. *Archives of Oral Biology* **21**, 393–400.

Buckley, M. J., Banes, A. J., Levin, L. G., Sumpio, B. E., Sato, M., Jordan, R., Gilbert, J., Link, G. & Tran Son Tay, R. (1988). Osteoblasts increase their rate of division and align in response to mechanical tension *in vitro*. *Bone and Mineral* **4**, 225–36.

Burgoyne, R. D., Morgan, A. & O'Sullivan, A. J. (1989). The control of cytoskeletal actin and exocytosis in intact and permeabilised chromaffin cells: role of calcium and protein kinase C. *Cell Signal* **1**, 323–34.

Greenberg, S., El Khoury, J., Di Virgilio, F., Kaplan, E. M. & Silver-stein, S. C. (1991). Ca²⁺-independent F-actin assembly and disassembly during Fc receptor-mediated phagocytosis in mouse macrophages. *Journal of Cell Biology* **113**, 757–67.

Hassell, T. M., Kimura, J. H. & Hascall, V. C. (1986). Proteoglycan core protein families. *Annual Review of Biochemistry* **55**, 539–67.

Heacock, C. S. & Bamburg, J. R. (1983). The quantitation of G- and F-actin in cultured cells. *Analytical Biochemistry* **135**, 22–36.

Hollenbeck, P. J., Bershadsky, A. D., Pletjushkina, O. Y., Tint, I. S. & Vasiliev, J. M. (1989). Intermediate filament collapse is an ATP-dependent and actin-dependent process. *Journal of Cell Biology* **92**, 621–31.

Howard, T. H. & Meyer, W. H. (1984). Chemotactic peptide modulation of actin assembly and locomotion in neutrophils. *Journal of Cell Biology* **98**, 1265–71.

Knowles, G. C. & McCulloch, C. A. G. (1992). Simultaneous localisation and quantitation of relative G- and F-actin content: optimisation of fluorescence labeling methods. *Journal of Histochemistry and Cytochemistry* **40**, 1605–12.

Koffer, A., Tatham, P. E. R. & Gomperts, B. D. (1990). Changes in the state of actin during the exocytotic reaction in mast cells. *Journal of Cell Biology* **111**, 919–27.

Kolega, J. (1986). Effects of mechanical tension on protrusive activity and microfilament and intermediate filament organisation in an epidermal epithelium moving in culture. *Journal of Cell Biology* **102**, 1400–11.

Korn, E. D. (1978). Biochemistry of actomyosin-dependent cell motility (a review). *Proceedings of the National Academy of Sciences USA* **75**, 588–99.

Melcher, A. H. & Chan, J. (1981). Phagocytosis and digestion of collagen by gingival fibroblasts *in vivo*, a study of serial sections. *Journal of Ultrastructural Research* **77**, 1–36.

Mochitate, K., Pawelek, P. & Grinnel, F. (1991). Stress relaxation of contracted collagen gels: disruption of actin filament bundles, release of cell surface fibronectin, and down-regulation of DNA and protein synthesis. *Experimental Cell Research* **193**, 198–207.

Murphy, G. & Reynolds, J. J. (1985). Current views on collagen degradation. Progress towards understanding the resorption of connective tissues. *Bioassays* **2**, 55–60.

Narayanan, A. S. & Page, R. C. (1983). Connective tissues of the periodontium: a summary of current work. *Collagen Related Research* **3**, 33–64.

Page, R. C. & Ammons, W. F. (1974). Collagen turnover in the gingiva and other mature connective tissues of the marmoset *Sanguinus oedipus*. *Archives of Oral Biology* **19**, 651–9.

Pender, N. & McCulloch, C. A. G. (1991). Quantitation of actin polymerisation in two human fibroblast sub-types responding to mechanical stretching. *Journal of Cell Science* **100**, 187–93.

Perrin, D., Moller, K., Hanke, K. & Soling, H.-D. (1992). cAMP and Ca^{2+}-mediated secretion in parotid acinar cells is associated with reversible changes in the organisation of the cytoskeleton. *Journal of Cell Biology* **116**, 127–34.

Samuels, R. H. A., Pender, N. & Last, K. S. (1993). The effects of orthodontic tooth movement on the glycosaminoglycan components of gingival crevicular fluid. *Journal of Clinical Periodontology* **20**, 371–7.

Sodek, J. (1977). A comparison of the rates of synthesis and turnover of collagen and non-collagenous protein in adult rat periodontal tissues and skin using a microassay. *Archives of Oral Biology* **22**, 655–65.

Sorsa, T., Ingman, T., Mikkoen, T., Suomalainen, K., Golub, I. M. & Thesleff, I. (1992). Characterisation of interstitial collagenase in gingival crevicular fluid during orthodontic tooth movement in man. In *Biological Mechanisms of Tooth Movement and Craniofacial Adaptation*, ed. Z. Davidovitch, pp. 47–54. Ohio State University.

Spudich, J. A. & Watt, S. (1971). The regulation of rabbit muscle contraction: biochemical studies of the interaction of the tropomyosin–troponin complex with actin and the proteolytic fragments of myosin. *Journal of Biological Chemistry* **246**, 4866–71.

Stossel, T. P. (1989). From signal to pseudopod. *Journal of Biological Chemistry* **264**, 18621–4.

Taylor, C. W. (1990). The role of G-proteins in transmembrane signalling. *Biochemical Journal* **272**, 1–13.

Ten Cate, A. R., Deporter, D. A. & Freeman, E. (1976). The role of fibroblasts in the remodelling of the periodontal ligament during physiological tooth movement. *American Journal of Orthodontics* **69**, 155–68.

Vitale, M. L., Rodriguez Del Castillo, A., Tchakarov, L. & Trifaro, J.-M. (1991). Cortical filamentous actin disassembly and scinderin redistribution during chromaffin cells stimulation precede exocytosis, a phenomenon not exhibited by gelsolin. *Journal of Cell Biology* **113**, 1057–67.

J. A. BEE, H.-X. LIU, N. CLARKE
and J. ABBOTT

Modulation of cartilage extracellular matrix turnover by pulsed electromagnetic fields (PEMF)

Introduction

The ability of low-level electrical stimulation to influence cell behaviour has aroused considerable attention in recent years. The presence and distribution of endogenous electrical currents during embryonic development (Jaffe, 1979) and regeneration (Borgens, Vanable & Jaffe, 1977) suggest that they may play important roles in directing cell movement and differentiation. However, it is the ability of skeletal tissues to respond to exogenous electrical stimulation that has attracted most interest. The demonstration that stressed bone possesses endogenous electrical potential and that exogenously applied electrical stimulation promotes osteogenesis *in vivo* (Yasuda, 1953; Bassett & Becker, 1962; Shamos, Lavine & Shamos, 1963) suggested that electrical stimulation might be employed to promote bone healing. Pulsed electromagnetic fields (PEMF) were subsequently shown to promote healing of fracture non-unions (Friedenberg, Harlow & Brighton, 1971) and are now widely applied for the successful clinical treatment of this condition in humans (Bassett, 1982; Bassett, Vades & Hernandez, 1982).

Although clinicians acknowledge that PEMF exert beneficial influences on damaged bone, the biological basis of this response has yet to be elucidated. How is exogenously applied PEMF stimulation converted into a cellular response which results in the restoration of normal tissue integrity? PEMF are generally considered to promote calcification in osteogenic tissue (Bassett *et al.*, 1979, 1982; Colacicco & Pilla, 1984) but this is also a complex process. For this reason, cell and tissue culture systems have been employed in an attempt to elucidate the mechanism by which PEMF modulate cell behaviour fundamental to tissue integrity. This approach eliminates potentially indirect effects on a tissue and

Society for Experimental Biology Seminar Series 54: *Biomechanics and Cells*, ed. F. Lyall & A. J. El Haj. © Cambridge University Press 1994, pp. 244–269.

focuses on the response of a defined targeted cell population. Electrical stimulation has thus been shown to promote actin polymerisation by fetal calvarial cells (Laub & Korenstein, 1984), possibly through a cyclic nucleotide- or calcium-dependent mechanism. Although analyses of the effect of PEMF on cAMP levels has produced conflicting and uncertain results (Norton, Rodan & Bourret, 1977; Davidovitch *et al.*, 1984), there is evidence to suggest that they may act together with factors such as parathyroid hormone to elevate cAMP content (Endo *et al.*, 1984; Hiraki *et al.*, 1987). Similar studies demonstrate that calcium uptake (Korenstein *et al.*, 1981), arachidonate release and prostaglandin E_2 synthesis (Johnson & Rodan, 1982; Davidovitch *et al.*, 1984) by isolated cells in culture are each increased in response to electrical stimulation. Nevertheless, the exact mechanism by which PEMF exert their effects on cell behaviour remains incompletely resolved. Equally intriguing is the overall response of a tissue to PEMF: damaged tissues tend to respond to PEMF preferentially with the result that normal structure and function is recovered more rapidly. It seems likely that PEMF enhance pre-existing biosynthetic and constructive pathways within a responsive cell population, a hypothesis supported by the fact that PEMF have not been observed to promote tissue disorganisation. Similarly, there is no evidence that PEMF evoke the appearance of abnormal cell phenotypes within a treated population.

It is now apparent that the *in vivo* and *in vitro* behaviour of a diverse range of differentiated cell types is influenced by exogenous electrical stimulation. Of particular interest is the ability of PEMF to modulate synthesis, and thus composition, of extracellular matrix. Cultured embryonic limb tissues respond by increasing collagen synthesis (Fitton-Jackson *et al.*, 1981; Fitton-Jackson & Bassett, 1981; Fitzsimmons *et al.*, 1986) while the synthesis of glycosaminoglycans is unaffected (Archer & Ratcliffe, 1983). Endochondral ossification in response to decalcified bone matrix *in vivo* is also enhanced by PEMF (Aaron, Ciombor & Jolly, 1989) indicating that they may stimulate chondrogenesis. Immature rabbit femoral articular cartilage responds *in vivo* by elevating its glycosaminoglycan content (Smith & Nagel, 1983) and numerous additional studies have demonstrated that extracellular matrix production by chondrocytes of diverse species and anatomical origin is modulated by PEMF treatment (reviewed by Brighton & McCluskey, 1986).

The physical properties of hyaline cartilage are conferred by the composition of its extracellular matrix (Heinegård & Oldberg, 1989). The principal components of this extracellular matrix are collagens, which confer structural rigidity, and proteoglycans, which confer compressional stiffness. The importance of proteoglycans to cartilage function is demonstrated by diseases, such as osteoarthritis, in which their content is abnormal (McDevitt & Muir, 1976; Hardingham & Bayliss, 1990). Since PEMF are known to influence extracellular matrix synthesis, might they also be applied to the treatment of cartilage disorders? Drawing conclusions from the numerous studies of the effect of PEMF on cartilage is difficult and potentially misleading. Different forms of PEMF have been employed for varying durations of treatment. The origin and age of the treated cartilage is equally diverse and includes chick sternum (Norton, 1982), rabbit articular (Smith & Nagel, 1983), bovine articular (Brighton, Unger & Stambough, 1984) and chick tibial (Bassett *et al.*, 1979). Although a number of studies have analysed the effect of PEMF on cultured cartilage, some of these have been performed with tissue explants (Brighton, Cronkey & Osterman, 1976; Fitton-Jackson & Bassett, 1981; Bee *et al.*, 1993), while others have utilised dissociated cells in monolayer (Norton, 1985; Norton & Rovetti, 1988; Sakai *et al.*, 1991). Consequently, different studies have reported quite different effects of PEMF on the synthesis of proteoglycans by cartilage. In some, proteoglycan synthesis is unaffected (Sakai *et al.*, 1991) while in others it is either reduced (Norton, 1982; Bee *et al.*, 1993) or elevated (Aaron & Plaas, 1987; Brighton *et al.*, 1984). PEMF have also been shown to reduce both the release and degradation of pre-existing sulphated glycosaminoglycans by cartilage explants (Bee *et al.*, 1993).

The studies described herein have focused on the response of a defined chondrocyte population cultured under standard conditions to a single form of PEMF. The effect of duration and frequency of treatment with this standard signal on DNA content, sulphated glycosaminoglycan synthesis, and sulphated glycosaminoglycan degradation by these standardised cartilage explants has been analysed. All experiments have been performed in the absence and presence of retinoic acid, a substance which is well known adversely to affect chondrocyte behaviour *in vitro* (Horton & Hassell, 1986; Horton, Yamada & Hassell, 1987; Benya, Brown & Padilla, 1988). Hypervitaminosis A also induces skeletal abnormalities *in vivo* (Clark & Seawright, 1968). The results demonstrate that the form of PEMF employed in these studies preserves the extracellular matrix integrity of the cartilage explants: although DNA content is unaffected and sulphated glycosaminoglycan synthesis is generally reduced, sulphated glycosaminoglycan degradation is virtually completely suppressed. The PEMF treatment regime at which this effect is maximal has also been identified.

Materials and methods

The present studies were all performed with embryonic chick sternal cartilage explanted to organ culture. Sternae were dissected from 16-day-old White Leghorn (*Gallus domesticus*) chick embryos and transferred to 35 mm plastic tissue culture dishes each containing 2 ml Ham's F12 nutrient medium supplemented with 10% fetal calf serum, 0.1 mg ml^{-1} ascorbate and antibiotics. In all experiments, each 35 mm tissue culture dish contained two identical sternae. Cultures were maintained in a humidified atmosphere at 37 °C under 5% CO_2 in air for the duration of the experiment. The culture dishes, three per stack, were supported on a platform between vertically placed coils (14 cm diameter, 13 cm intercoil distance) mounted on a plastic framework (Electro-Biology, Inc., Parsippany, New Jersey, USA; Fig. 1). The PEMF signal employed in the present studies comprised +0.7, −0.5, −0.7 and +0.5 mV, respectively, with a cycle duration of 28 ms (Fig. 2). Experimental cultures were treated with PEMF according to one of five different daily application regimes: (i) 1 h on:23 h off, (ii) 3 h on:21 h off, (iii) a repeating cycle of 1 h on:1 h off, (iv) a repeating cycle of 3 h on:3 h off, and (v) a repeating cycle of 12 h on:12 h off (Fig. 3). Control cultures were maintained

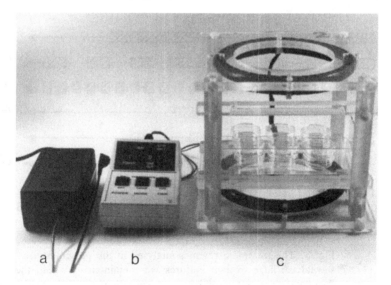

Fig. 1. Apparatus constructed for the treatment of chick sternal cartilage explant cultures with pulsed electromagnetic fields. The system comprises a battery charger (a) connected to a power unit with built-in timer facility (b). Cultures were supported between vertically placed coils mounted on a plastic framework (c).

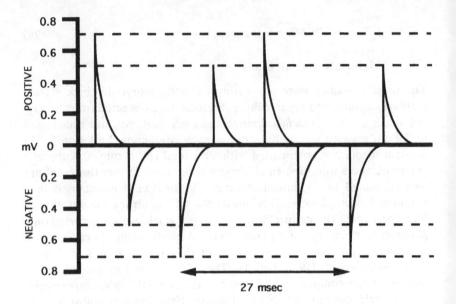

Fig. 2. Schematic representation of the pulsed electromagnetic field used in the present study. Sternal cartilage explants were maintained in culture for up to 48 h during which they were exposed to this standard signal for various durations and cycles of treatment.

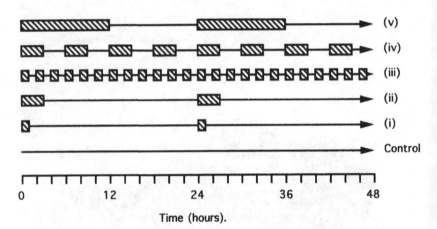

Fig. 3. The treatment regimes analysed in the present study. Control sternal cartilage explant cultures were maintained within the plastic framework without receiving electrical stimulation (Control). Experimental cultures maintained under identical conditions were exposed to PEMF for 1 h per day (i), 3 h per day (ii), a repeating cycle of 1 h on:1 h off (iii), a repeating cycle of 3 h on:3 h off (iv), or a repeating cycle of 12 h on:12 h off (v). The hatched boxes represent the duration for which cultures were exposed to PEMF.

within the coils under identical environmental conditions but in the continuous absence of PEMF. In parallel experiments performed under identical conditions the nutrient medium was supplemented with 10^{-5} M all *trans*-retinoic acid (Sigma Chem. Co., Poole, Dorset).

Analysis of DNA content

The DNA content of sternae was determined immediately after their dissection from the embryo and following explant culture in either the absence or presence of PEMF. Each sample comprised two sternae, either freshly isolated or cultured, and all experiments were performed at least in triplicate. To each sample was added 0.7 ml 50 mM Tris, pH 8.0:100 mM, EDTA:100 mM NaCl:1% SDS and 250 μg proteinase K (Boehringer Mannheim UK). Following incubation at 55 °C overnight, each sample received 0.7 ml phenol:chloroform (1:1, v/v), was shaken for 15 min, and then spun for 10 min in a microfuge. The upper phase of each sample was transferred to a fresh tube, received 0.7 ml of the phenol:chloroform mixture, was shaken for 15 min, and spun for 10 min in a microfuge. The resultant aqueous phase was extracted once with an equal volume (0.7 ml) chloroform, shaken for 15 min, and spun for 5 min in a microfuge. The upper phase was transferred to a fresh tube, received 0.6 volumes isopropanol, was shaken vigorously and then microfuged for 5 min. The supernatant was discarded and the visible DNA pellet washed with 1 ml 70% ethanol, recovered by microfuge centrifugation, washed with 1 ml 100% ethanol, recovered by microfuge centrifugation, and then allowed to air dry. Each sample received 100 μl 10 mM Tris HCl, pH 8.0:1 mM EDTA, pH 8.0, and was then incubated overnight to dissolve the DNA. 20 μl aliquots from each sample were diluted to a final volume of 1 ml with 10 mM Tris HCl, pH 8.0:1 mM EDTA, pH 8.0 and the optical density at both 260 and 280 nm determined spectrophotometrically. The purity of each preparation was determined from the ratio of these two measurements and in the present study was never less than 1.8. Total DNA content for each sample was calculated on the basis that 1 optical density unit at 260 nm is equivalent to 50 μg DNA ml^{-1}.

The effect of PEMF treatment on DNA content

The data presented in Fig. 4 demonstrate the DNA content of explants cultured in either the absence of (*a*) or presence (*b*) of 10^{-5} M retinoic acid. In the absence of retinoic acid, the DNA content of control explants cultured in the absence of PEMF remains relatively constant during the course of the experiment. Compared with explants at the onset of the experiment (Fig. 4*a*, 0 [0]), the DNA content of control explants does not change significantly after either 24 (Fig. 4*a*, 24 [0]) or 48 h in culture

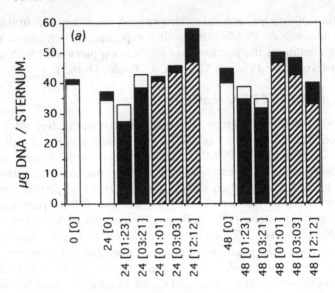

HOURS IN CULTURE.

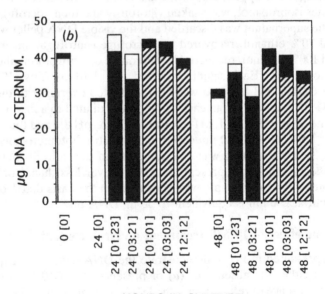

HOURS IN CULTURE.

(Fig. 4*a*, 48 [0]). The different PEMF treatment regimes demonstrate subtle effects on the DNA content of parallel cultures. In general, exposure for either 1 (Fig. 4*a*, 24 [01:23] and 48 [01:23]) or 3 h per day (Fig. 4*a*, 24 [03:21] and 48 [03:21]) tends to cause a slight reduction in DNA content. Repeating cycles of 1 h on:1 h off (Fig. 4*a*, 24 [01:01] and 48 [01:01]), 3 h on:3 h off (Fig. 4*a*, 24 [03:03] and 48 [03:03]), and 12 h on:12 h off (Fig. 4*a*, 24 [12:12] and 48 [12:12]) tend to cause a slight increase in DNA content. However, the data for each of the PEMF treatments are not significantly different from each other or from controls. Thus, throughout the culture period the DNA content of explants is not significantly affected by any of the PEMF treatments.

In the presence of 10^{-5} M retinoic acid (Fig. 4*b*) the DNA content of explants cultured in the absence of PEMF decreases markedly after 24 h (Fig. 4*b*, 24 [0]) and remains at this reduced level after 48 h (Fig. 4*b*, 48 [0]). In parallel cultures exposed to PEMF the DNA content after 24 h remains similar to that at the onset of culture and is consistently greater than that of untreated controls. After 24 h four of the PEMF treatments – 1 h per day and repeating cycles 1 h on:1 h off, 3 h on:3 h off and 12 h on:12 h off – are not significantly different from one another or from the starting value but are significantly greater than untreated explants (Fig. 4*b*). Only one of the PEMF treatments – 3 h per day (Fig. 4*b*, 24 [03:21]) – is not significantly different from the untreated explants. After 48 h only two of the PEMF treatments, 1 h per day (Fig. 4*b*, 48 [01:23]) and a repeating cycle of 1 h on:1 h off (Fig. 4*b*, 48 [01:01]), maintain a DNA content significantly greater than untreated controls (Fig. 4*b*, 48 [0]).

These data demonstrate that the DNA content of sternal cartilage explants is not adversely affected by any of the PEMF treatment regimes analysed. Similar results have been reported elsewhere (Archer & Ratcliffe, 1983; Norton, 1985; Sakai *et al.*, 1991). In general, PEMF-treated explants tend to exhibit a slight increase in DNA content compared with

---

Fig. 4. The effect of PEMF treatment on DNA content. The DNA content of sternal cartilage explants was determined at the onset of (0 [0]) and after 24 (24 [0]) and 48 h (48 [0]) culture in the absence (*a*) or presence (*b*) of 10^{-5} M retinoic acid. In both the absence and presence of retinoic acid, the effect of PEMF on the DNA content of explants was determined after 24 and 48 h treatment according to each of the application regimes ([01:23], [03:21], [01:01], [03:03], and [12:12]). Columns are plotted to the average of the data with each cap representing the positive standard deviation.

parallel controls. Retinoic acid causes a marked reduction in DNA content and this effect is significantly prevented by PEMF treatment. The more pronounced effect of PEMF on DNA content in the presence of retinoic acid probably reflects the enhanced responsiveness to PEMF by explants exposed to adverse environmental conditions. Although cell proliferation has not been determined but may also be stimulated by PEMF (Norton, 1985; Sakai *et al.*, 1991) the data presented indicate that PEMF preserve DNA content in cultured cartilage explants.

Analysis of radiolabelled sulphated glycosaminoglycan content

The effect of PEMF treatment on both the synthesis and degradation of sulphated glycosaminoglycans by sternal cartilage explants has been analysed. To determine the effect of PEMF treatment on the synthesis of sulphated glycosaminoglycans, carrier-free [^{35}S]sulphuric acid (Amersham International, Amersham) was added to control and experimental cultures at a final concentration of 5 μCi ml^{-1} after 0 and 24 h of culture. Twenty-four hours after the addition of [^{35}S]sulphuric acid, equivalent to 24 and 48 h of culture, sternal explants and nutrient media were harvested separately and analysed for the incorporation of [^{35}S]sulphate into sulphated glycosaminoglycans. The effect of PEMF $\pm$ retinoic acid on the incorporation of [^{35}S]sulphate into sulphated glycosaminoglycans either retained by the explant or released into the medium was determined after 24 and 48 h of culture.

To determine the effect of PEMF treatment on sulphated glycosaminoglycan degradation, 10 μCi [^{35}S]sulphuric acid in 100 μl distilled water was administered onto the chorioallantoic membrane through a small hole in the shell on the sixth day of embryonic development. Following the administration of isotope, the hole in the shell was sealed with adhesive tape and eggs returned to the incubator to facilitate continued development. Biosynthetically radiolabelled sternae were isolated on embryonic day 16 and either immediately analysed for the incorporation of [^{35}S]sulphate into sulphated glycosaminoglycans or cultured as explants. In either the absence or presence of retinoic acid, cultures were randomly divided into control and experimental (PEMF-treated) groups. Sternal explants and nutrient media were harvested separately after 24 and 48 h of culture and subjected to analysis of [^{35}S]sulphate labelled glycosaminoglycan content.

A minimum of three parallel cultures, i.e. two sternae per sample, were analysed for each treatment and for each incubation period. To each sample, comprising two isolated sternal explants in 1 ml distilled water or the total 2 ml nutrient medium, was added 250 μg whale and shark cartilage mixed-isomer chondroitin sulphates (Sigma Chem. Co.) and 250 μg human umbilical cord hyaluronic acid (Sigma Chem. Co.) as carriers. Concentrated Tris HCl was added to achieve a final concentration of 0.2 M (pH 8.0). Samples were immersed in a boiling water bath for 5 min, cooled, and then incubated for 48 h at 50 °C with predigested *Streptomyces griseus* pronase (0.18 mg ml^{-1}, Boehringer Mannheim UK, Lewes, East Sussex) in the presence of $CaCl_2$ and ethanol (de la Haba & Holtzer, 1965). Fresh enzyme was added after 24 h. Upon completion of the pronase digestion, each sample received 10 ml distilled water and was then applied to a column of freshly regenerated Dowex 1-Cl-resin (0.5 × 6.0 cm; AG1-X2 [200–400 mesh; Bio-Rad Laboratories Ltd., Watford, Herts.]). Columns were eluted using a stepwise gradient of increasing NaCl molarity (Schiller, Slover & Dorfman, 1961; Conrad & Dorfman, 1974; Conrad, Hamilton & Haynes, 1977). Fractions of 10 ml were collected and aliquots of 0.5 ml were analysed for radioactivity using an LKB 1214 Rackbeta Liquid Scintillation Counter and an aqueous counting scintillant (Amersham International). Unincorporated [^{35}S]sulphate was eluted with a total volume of 60 ml 0.2 M NaCl. The columns were thereafter eluted with 10 ml each of 0.3, 0.4, 0.5 and 1.0 M NaCl. The bulk of the sulphated glycosaminoglycans was then eluted with 30 ml 2.0 M NaCl; incorporation of [^{35}S]sulphate into this fraction was used in this study. The synthesis of sulphated glycosaminoglycans by sternae *in vitro* and their retention by the explant or release into the nutrient medium was determined by the total amount of radioactivity eluting in the 2.0 M NaCl fraction. All data were expressed per μg DNA. The incorporation of [^{35}S]sulphate into sulphated glycosaminoglycans by sternae *in ovo* was determined immediately following their isolation from the embryo and expressed with reference to DNA content as an initial value of 100%. Sulphated glycosaminoglycans biosynthetically radio-labelled *in ovo* by corresponding, matched sternae and then recovered from either the explant or the nutrient medium after the culture period was expressed with reference to DNA content as a percentage of the initial value. At each time point within each experiment, total incorporation or percentage recovery of sulphated glycosaminoglycans was expressed as the mean ± SD.

The effect of PEMF on sulphated glycosaminoglycan synthesis

The incorporation of [^{35}S]sulphate into sulphated glycosaminoglycans by sternal cartilage explants cultured in the absence of retinoic acid is

254   J. A. BEE *et al.*

presented in Fig. 5. The effect of PEMF treatment on the recovery of radiolabelled sulphated glycosaminoglycans either retained by the explant (Fig. 5a) or released into the nutrient medium (Fig. 5b) has been determined after 24 and 48 h of culture.

In the absence of PEMF, the incorporation of [^{35}S]sulphate into sulphated glycosaminoglycans retained by the explant is relatively constant after both 24 (Fig. 5a, 24 [0]) and 48 h (Fig. 5a, 48 [0]). None of the PEMF treatments analysed exert a statistically significant effect during the first 24 h of culture. However, after 48 h all PEMF treatments significantly reduce the incorporation of [^{35}S]sulphate into sulphated glycosaminoglycans. Although each of the PEMF treatment regimes exert subtly different effects, a general trend is apparent. Exposure periods of 1 h, either per day (Fig. 5a, 48 [01:23]) or on a repeating 1 h on:1 h off cycle (Fig. 5a, 48 [01:01]) exert the greatest effect on the reduction of [^{35}S]sulphate incorporation into sulphated glycosaminoglycans retained by the explant. Increasing the duration of the exposure period slightly reduces the effect which is least with a repeating 12 h on:12 h off cycle (Fig. 5a, 48 [12:12]).

PEMF exert similar effects on the recovery of radiolabelled sulphated glycosaminoglycans from the nutrient medium (Fig. 5b). These data reflect the incorporation of [^{35}S]sulphate into sulphated glycosaminoglycans which are then released from the explant. After 24 h, compared with controls (Fig. 5b, 24 [0]) none of the PEMF treatments exert a significant effect although in cultures treated for 1 h per day (Fig. 5b, 24[01:23]) there is a slight elevation. After 48 h all PEMF treatments tend to reduce the levels of radiolabelled sulphated glycosaminoglycans recovered with the nutrient medium. The most significant effect is observed in cultures treated with a repeating 1 h on:1 h off cycle (Fig. 5b, 48 [01:01]). Increasing the duration of the exposure period, on either a daily or a repeating cycle, correlates directly with an increase in the release of radiolabelled sulphated glycosaminoglycans into the nutrient medium.

Parallel experiments also were performed in the presence of 10^{-5} M

Fig. 5. The effect of PEMF treatment on the incorporation of [^{35}S]sulphate into sulphated glycosaminoglycans. Sternal cartilage explants were maintained in culture for 24 and 48 h. Incorporation of [^{35}S]H$_2$SO$_4$ into sulphated glycosaminoglycans either retained by the explant (a) or released into the medium (b) was determined after 24 and 48 h culture in the absence ([0]) or presence of each of the PEMF treatment regimes ([01:23], [03:21], [01:01], [03:03], and [12:12]). Columns are plotted to the average of the data with each cap representing the positive standard deviation.

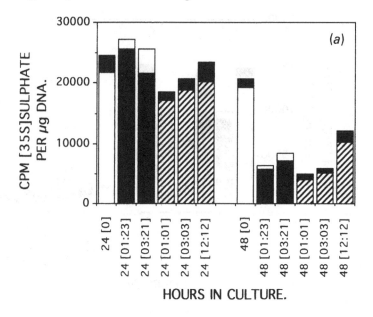

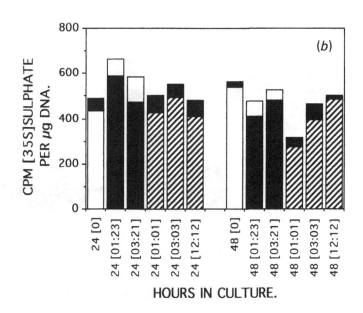

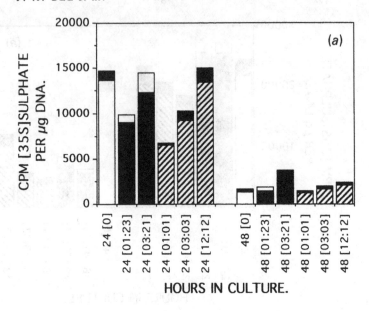

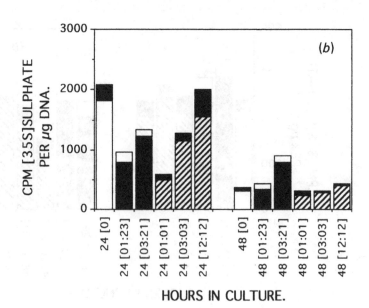

retinoic acid (Fig. 6). Under these conditions, the influence of the PEMF signal employed in these studies on the incorporation of [^{35}S]sulphate into sulphated glycosaminoglycans recovered with the explant (Fig. 6*a*) or the nutrient medium (Fig. 6*b*) is especially pronounced. Correlation between treatment regime and effect is also more apparent. After 24 h, explants treated for 3 h per day (Fig. 6*a*, 24 [03:21]) or a repeating cycle of 12 h on:12 h off (Fig. 6*a*, 24 [12:12]) are not significantly different from controls. However, in cultures treated for 1 h per day (Fig. 6*a*, 24 [01:23]) or repeating cycles of 1 h on:1 h off (Fig. 6*a*, 24 [01:01]) and 3 h on:3 h off (Fig. 6*a*, 24 [03:03]) the incorporation of [^{35}S]sulphate into sulphated glycosaminoglycans recovered with the explant is substantially reduced. This effect is directly proportional to the duration of the exposure period. As the duration of the exposure period increases the incorporation of [^{35}S]sulphate into sulphated glycosaminoglycans retained by the explant increases towards the levels demonstrated by untreated controls. It is worthy of note that although all cultures exposed to repeating cycles of PEMF received a total of 12 h treatment per day, the three different treatment regimes produced strikingly different effects. This same pattern persisted after 48 h although the incorporation of [^{35}S]sulphate into sulphated glycosaminoglycans retained by both control and treated explants was greatly reduced. Consequently, with the exception of 3 h exposure per day (Fig. 6*a*, 48 [03:21]) which increases the incorporation of [^{35}S]sulphate into sulphated glycosaminoglycans, none of the PEMF treatments was significantly different from controls (Fig. 6*a*, 48 [0]).

An identical pattern of response is observed with the levels of radio-labelled sulphated glycosaminoglycans recovered with the nutrient medium after both 24 and 48 h (Fig. 6*b*). All PEMF treatment regimes reduce the levels of radiolabelled sulphated glycosaminoglycans recovered with the nutrient medium after 24 h. This effect also increases with decreasing duration of the exposure period. After 48 h, only cultures treated for 3 h on:21 h off (Fig. 6*b*, 48 [03:21]) are significantly different

Fig. 6. The effect of 10^{-5} M retinoic acid $\pm$ PEMF treatment on the incorporation of [^{35}S]sulphate into sulphated glycosaminoglycans. Sternal cartilage explants were maintained in culture for 24 and 48 h in the presence of retinoic acid. Incorporation of [^{35}S]H$_2$SO$_4$ into sulphated glycosaminoglycans either retained by the explant (*a*) or released into the medium (*b*) was determined after 24 and 48 h culture in the absence ([0]) or presence of each of the PEMF treatment regimes ([01:23], [03:21], [01:01], [03:03], and [12:12]). Columns are plotted to the average of the data with each cap representing the positive standard deviation.

from controls (Fig. 6*b*, 48 [0]) and demonstrate an elevation in radio-labelled sulphated glycosaminoglycans released by the explants.

Proteoglycans represent a major component of hyaline cartilage extracellular matrix. The data presented herein demonstrate that the incorporation of [^{35}S]sulphate into sulphated glycosaminoglycans, indicative of proteoglycan synthesis, by sternal cartilage explanted to culture is influenced by PEMF. In addition, PEMF treatment affects the levels of newly synthesised sulphated glycosaminoglycans either retained by the explant or released to the nutrient medium. Similar effects of PEMF on cartilage have been reported elsewhere although they have not been analysed in detail (Farndale, 1982; Smith, 1984). PEMF treatment reduces both the incorporation of [^{35}S]sulphate into sulphated glycosaminoglycans and the release of newly synthesised sulphated glycosaminoglycans from explants. The extent of this response is a consistent reflection of the PEMF treatment regime. The influence of PEMF on sulphated glycosaminoglycan synthesis and the fate of newly synthesised material is apparent by 48 h in the absence of retinoic acid and within 24 h in its presence. Retinoic acid therefore appears to enhance the responsiveness of cartilage explants to PEMF. The relationship between treatment regime and response is extremely consistent and demonstrates a general pattern in the absence of retinoic acid. This dose-related response is both identical and highly significant in the presence of retinoic acid. Regardless of culture conditions, short periods of PEMF treatment exert the greatest effects. The data also demonstrate that within a 24 h period the response of cultured cartilage explants to PEMF directly reflects the application regime but is independent of the overall duration of treatment.

The effect of PEMF on sulphated glycosaminoglycan degradation

[^{35}S]sulphuric acid was administered *in ovo* on the sixth day of embryonic development and radiolabelled sternae isolated 10 days later. Sternae were explanted to culture and the fate of pre-existing sulphated glycosaminoglycans determined (Fig. 7). The levels of [^{35}S]sulphated glycosaminoglycans were determined immediately after isolation of the sternae and, relative to DNA content, expressed as a starting value 100%. Equivalently radiolabelled sternae were explanted to culture in either the absence or presence of PEMF. The recovery of [^{35}S]sulphated glycosaminoglycans with either the explant (Fig. 7*a*) or nutrient medium (Fig. 7*b*) was determined after 24 and 48 h. Relative to DNA content, the recovery of [^{35}S]sulphated glycosaminoglycans was then expressed as a percentage of the 100% starting value.

In the absence of PEMF, the recovery of [^{35}S]sulphated glycosaminoglycans with the explant remains relatively constant after 24 h (Fig. 7a, 24 [0]) but decreases substantially by 48 h (Fig. 7a, 48 [0]). Over the same period, the recovery of [^{35}S]sulphated glycosaminoglycans with the nutrient medium (Fig. 7b, 24 [0] and 48 [0]) remains consistently low. Thus, this reduction in content is not due to a release of [^{35}S]sulphated glycosaminoglycans from the explant to the nutrient medium. PEMF treatment according to any of the five regimes employed dramatically affects the recovery of [^{35}S]sulphated glycosaminoglycans. After 24 h the recovery of [^{35}S]sulphated glycosaminoglycans with the explant is only significantly different from controls with two of the PEMF treatments. Cultures exposed to repeating cycles of either 12 h on:12 h off (Fig. 7b, 24 [12:12]) or 3 h on:3 h off (Fig. 7a, 24 [03:03]) exhibit a slight reduction in [^{35}S]sulphated glycosaminoglycan content. Simultaneously, each of the PEMF treatments results in a slight elevation in the levels of [^{35}S]sulphated glycosaminoglycans recovered with the nutrient medium (Fig. 7b, 24).

After 48 h, treatment with PEMF exerts striking effects on the recovery of [^{35}S]sulphated glycosaminoglycans. Compared with controls, all PEMF treatments cause a significant increase in the proportion of pre-existing sulphated glycosaminoglycans recovered with the explant (Fig. 7a). The proportion of pre-existing sulphated glycosaminoglycans recovered with the explant also correlates strongly with the PEMF treatment regime. Repeating cycles of 3 h on:3 h off (Fig. 7a, 48 [03:03]), 1 h on:1 h off (Fig. 7a, 48 [01:01]) and 12 h on: 12 h off (Fig. 7a, 48 [12:12]) are increasingly effective. The most effective repeating cycle, 12 h on:12 h off, or a single treatment of 1 h per day (Fig. 7a, 48 [01:23]) both result in an almost complete recovery of pre-existing sulphated glycosaminoglycans with the explant. The levels of [^{35}S]sulphated glycosaminoglycans recovered with the explant after a single PEMF treatment of 3 h per day (Fig. 7a, 48 [03:21]) are actually greater than at the onset of the culture period. This experiment has been repeated several times and these results are extremely consistent. This result possibly reflects a stimulation of sulphated glycosaminoglycan synthesis utilising a pre-existing pool of precursors.

The levels of [^{35}S]sulphated glycosaminoglycans recovered with the nutrient medium continue to be elevated by PEMF treatment after 48 h (Fig. 7b, 48). Three of the PEMF treatment regimes, 1 h per day (Fig. 7b, 48 [01:23]) and repeating cycles of 1 h on:1 h off (Fig. 7b, 48 [01:01]) and 3 h on:3 h off (Fig. 7b, 48 [03:03]), do not differ in their effect. Treatment for 3 or 12 h per day cause a sequential increase in the levels of [^{35}S]sulphated glycosaminoglycans recovered with the nutrient medium.

Sternae biosynthetically radiolabelled *in ovo* with [^{35}S]H$_2$SO$_4$ were also

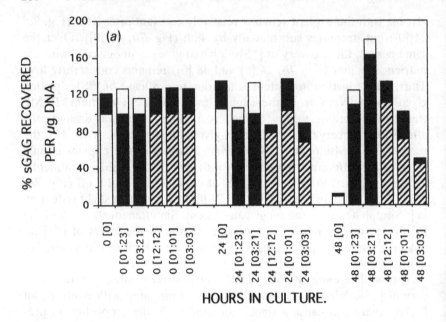

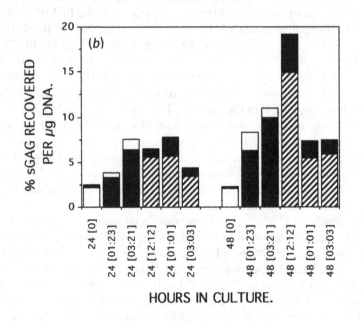

cultured in the presence of 10^{-5} M retinoic acid (Fig. 8). [^{35}S]sulphated glycosaminoglycan content was determined immediately following isolation of the sternae and, relative to DNA content, expressed as an initial value of 100%. The effect of treatment with PEMF according to each of the application regimes on the recovery of [^{35}S]sulphated glycosaminoglycans with either the explant (Fig. 8a) or the nutrient medium (Fig. 8b) was then determined after 24 and 48 h of culture.

Compared with cultures maintained in its absence (Fig. 7), the presence of retinoic acid significantly elevates the levels of pre-existing [^{35}S]sulphated glycosaminoglycans that are both recovered with the explant (Fig. 8a) and released to the medium (Fig. 8b). In the absence of PEMF, approximately 80% of the pre-existing [^{35}S]sulphated glycosaminoglycans are recovered with the explant after both 24 (Fig. 8a, 24 [0]) and 48 h (Fig. 8a, 48 [0]). This reduction in [^{35}S]sulphated glycosaminoglycan content is largely due to their release from the explant. Thus, approximately 10% is recovered with the nutrient medium after 24 h (Fig. 8b, 24 [0]) and 20% after 48 h (Fig. 8b, 48 [0]). These data indicate that 10^{-5} M retinoic acid significantly reduces the degradation of pre-existing sulphated glycosaminoglycans.

PEMF treatment in the presence of retinoic acid affects the levels of pre-existing [^{35}S]sulphated glycosaminoglycans recovered with either the explant or the nutrient medium depending upon the application regime. After 24 h in culture, treatment for 1 h per day (Fig. 8a, 24 [01:23]) or continuous cycles of 1 h on:1 h off (Fig. 8a, 24 [01:01]) and 3 h on:3 h off (Fig. 8a, 24 [03:03]) are not significantly different from untreated controls in the recovery of [^{35}S]sulphated glycosaminoglycans with the explant. However, following treatment for 3 h per day (Fig. 8a, 24 [03:21]) or a continuous cycle of 12 h on:12 h off (Fig. 8a, 24 [12:12]) the levels which are recovered with the explant are significantly elevated relative to controls (Fig. 8a, 24 [0]). Analysis of [^{35}S]sulphated glycosaminoglycans recovered with the nutrient medium after 24 h (Fig. 8b, 24)

Fig. 7. The effect of PEMF on the recovery of pre-existing sulphated glycosaminoglycans. Sternae were isolated from embryos previously administered with [^{35}S]H$_2$SO$_4$. [^{35}S]sulphated glycosaminoglycan content was determined immediately (a, 0) and following explant culture. The recovery of [^{35}S]sulphated glycosaminoglycans with the cartilage explant (a) and medium (b) was determined after 24 and 48 h culture in the absence ([0]) or presence of each of the PEMF treatment regimes ([01:23], [03:21], [01:01], [03:03], and [12:12]). Columns are plotted to the average of the data with each cap representing the positive standard deviation.

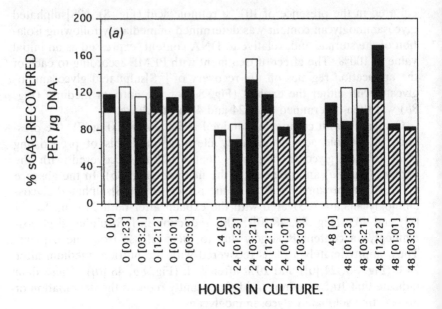

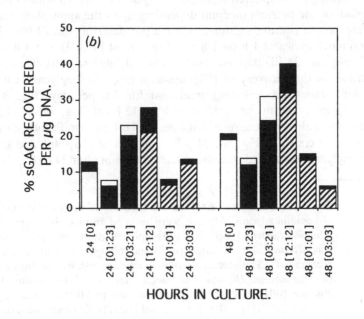

demonstrates that different PEMF treatments have strikingly different effects. A repeating cycle of 3 h on:3 h off (Fig. 8*b* 24 [03:03]) is not significantly different from untreated controls (Fig. 8*b*, 24 [0]). Treatment for 1 h per day (Fig. 8*b*, 24 [01:23]) or a repeating cycle of 1 h on:1 h off (Fig. 8*b*, 24 [01:01]) reduces the levels of [^{35}S]sulphated glycosamino-glycans recovered with the nutrient medium. In contrast, treatment for 3 h per day (Fig. 8*b*, 24 [03:21]) or a repeating cycle of 12 h on:12 h off (Fig. 8*b*, 24 [12:12]) elevates the levels of [^{35}S]sulphated glycosamino-glycans recovered with the nutrient medium.

Similar effects are observed after 48 h treatment of cultures with PEMF. The levels of [^{35}S]sulphated glycosaminoglycans recovered with the explant (Fig. 8*a*, 48) are not significantly different from controls (Fig. 8*a*, 48 [0]) with any of the five PEMF treatment regimes. However, levels do tend to be slightly higher with treatment for either 3 h per day (Fig. 8*a*, 48 [03:21]) and a repeating cycle of 12 h on:12 h off (Fig. 8*a*, 48 [12:12]). Nevertheless, PEMF treatment continues significantly to affect the levels of [^{35}S]sulphated glycosaminoglycans recovered with the nutri-ent medium (Fig. 8*b*, 48). The overall pattern of response to each of the treatment regimes after 48 h parallels the effects demonstrated after 24 h. The only exception is treatment on a repeating cycle of 3 h on:3 h off (Fig. 8*b*, 48 [03:03]) which now significantly reduces the levels of [^{35}S]sulphated glycosaminoglycans recovered with the nutrient medium.

Analysis of exogenous factors which influence the integrity of mature cartilage tends to focus on their effects on the synthesis of defined com-ponents of the extracellular matrix. Nevertheless, their effect on the extracellular matrix synthesised and assembled prior to treatment is equally important. The developing chick embryo is an excellent model with which to perform this latter type of investigation. Radioisotopes can be administered during the early stages of embryonic development to produce tissues that are highly biosynthetically labelled. Employing such

---

Fig. 8. The effect of 10^{-5} M retinoic acid ± PEMF on the recovery of pre-existing sulphated glycosaminoglycans. Sternae were isolated from embryos previously administered with [^{35}S]H$_2$SO$_4$. [^{35}S]sulphated glycos-aminoglycan content was determined immediately (*a*, 0) and following explant culture in the presence of retinoic acid. The recovery of [^{35}S]sul-phated glycosaminoglycans with the cartilage explant (*a*) and medium (*b*) was determined after 24 and 48 h culture in the absence ([0]) or presence of each of the PEMF treatment regimes ([01:23], [03:21], [01:01], [03:03], and [12:12]). Columns are plotted to the average of the data with each cap representing the positive standard deviation.

an approach with the developing sternum, the present studies demonstrate a profound effect of PEMF on the fate of pre-existing sulphated glycosaminoglycans *in vitro*. PEMF treatment preserves pre-existing sulphated glycosaminoglycan content and retention by sternal cartilage explants. This effect is strikingly influenced by the treatment regime with 3 h per day being the most effective in either the absence of presence of retinoic acid. PEMF treatment also influences the levels or pre-existing sulphated glycosaminoglycans that are released from the explant and recovered with the nutrient medium. Release of pre-existing sulphated glycosaminoglycans from the explant is elevated by retinoic acid and this effect can be significantly reduced by the additional application of PEMF treatment.

Conclusions

The biosynthetic behaviour of embryonic chick sternal cartilage explanted to culture shows many similarities with the *in vivo* development of cartilage disorders, such as osteoarthritis. In experimentally induced osteoarthritis, an initial elevation of sulphated proteoglycan synthesis (Hardingham & Bayliss, 1990) is followed by their excessive degradation (Pelletier *et al.*, 1992), leading to a progressive loss of articular cartilage structure and function (McDevitt & Muir, 1976). Thus, factors which can modulate abnormal levels of sulphated proteoglycan synthesis and/or degradation could potentially be employed to conserve normal cartilage integrity.

The studies reported herein describe the response of intact cartilage explants to PEMF. Explants were employed since they would be expected to exhibit a more representative response than dispersed cells in monolayer culture. Explant culture also alleviates potentially abnormal chondrocyte biosynthetic behaviour resulting from enzymatic dissociation and subsequent monolayer culture (Holtzer *et al.*, 1960).

Embryonic chick sternal cartilage explanted to culture initially exhibits elevated synthesis of sulphated glycosaminoglycans. This level of sulphated glycosaminoglycan synthesis declines sharply after 72 h in culture (Bee *et al.*, 1993). After the initial elevation of their synthesis, degradation of sulphated glycosaminoglycans is increased. Both the initial elevation of sulphated glycosaminoglycan synthesis and subsequent increased degradation are reduced by 10^{-5} M retinoic acid. These data suggest that sulphated glycosaminoglycan synthesis is abnormally elevated during the initial stages of explant culture. As a consequence of this, degradation of pre-existing sulphated glycosaminoglycans is increased in an attempt to restore tissue integrity. Both newly synthesised and pre-existing sulphated glycosaminoglycans are released from the explant to the nutrient medium and this is increased by retinoic acid.

The present study demonstrates that PEMF are capable of modulating the synthesis, degradation and release of sulphated glycosaminoglycans by embryonic chick sternal cartilage explanted to culture. PEMF treatment reduces both the initial elevation of sulphated glycosaminoglycan synthesis and subsequent increase in their degradation. Degradation of pre-existing sulphated glycosaminoglycans is greatly reduced in the presence of PEMF. It is therefore unlikely that the reduction in apparent levels of sulphated glycosaminoglycans newly synthesised in the presence of PEMF are actually higher but are decreased due to increased degradation. PEMF also reduce the levels of sulphated glycosaminoglycans that are released from the explant to be recovered with the nutrient medium. This effect is especially significant in the presence of retinoic acid. Each of these effects is influenced by the application regime. Short periods of exposure to PEMF, most notably 3 h per day, are clearly the most effective in maintaining a more normal pattern of sulphated glycosaminoglycan turnover. In addition, the effect of three different repeating cycles of exposure has been analysed. Although the total duration of exposure was 12 h per day, the overall response clearly reflects the duration of an individual treatment period.

The PEMF employed in the present study has previously been shown to conserve the histological structure of embryonic chick sternal cartilage explanted to culture (Bee *et al.*, 1993). The identification of the optimal PEMF treatment regime producing the maximum beneficial response should now enable the molecular mechanism by which cartilage responds to low level electrical stimulation to be defined. PEMF clearly have considerable therapeutic potential for the treatment of cartilage disorders. It is now appropriate that their effect on damaged or diseased cartilage *in vivo* should be analysed.

References

Aaron, R. K., Ciombor, D. McK. & Jolly G. (1989). Stimulation of experimental endochondral ossification by low-energy pulsing electromagnetic fields. *Journal of Bone and Mineral Research* **4**, 227–31.

Aaron, R. K. & Plaas, A. H. K. (1987). Stimulation of proteoglycan synthesis in articular chondrocyte cultures by a pulsed electromagnetic field. *Transactions of the Orthopaedic Research Society* **12**, 273.

Archer, C. W. & Ratcliffe, N. A. (1983). The effects of pulsed magnetic fields on chick embryo cartilaginous skeletal rudiments *in vitro*. *Journal of Experimental Zoology* **225**, 243–56.

Bassett, C. (1982). Pulsing electromagnetic fields: a new method to modify cell behavior in calcified and non-calcified tissues. *Calcified Tissue International* **34**, 1–8.

Bassett, C. & Becker, R. O. (1962). Generation of electrical potential by bone in response to mechanical stress. *Science* **137**, 1063–4.

Bassett, C., Choski, H., Hernandez, E., Pawluk, R. & Strop, M. (1979). The effect of pulsing electromagnetic fields on cellular calcium and calcification of non-unions. In *Electrical Properties of Bone and Cartilage*, ed. C. T. Brighton, J. Black & S. Pollack, pp. 427–42. New York: Grune and Stratton.

Bassett, C., Vades, M. & Hernandez, E. (1982). Modification of fracture repair with selected pulsing electromagnetic fields. *Journal of Bone and Joint Surgery* **64**, 888–95.

Bee, J. A., Clarke, N., Wesse, M., Hawkins, P. A. & Abbott, J. (1993). The effect of pulsed electromagnetic fields on sulfated glycosaminoglycan turnover by chick sternal cartilage explanted *in vitro*. *Journal of Orthopaedic Research* (in press).

Benya, P. D., Brown, P. D. & Padilla, S. R. (1988). Microfilament modification by dihydrocytochalasin B causes retinoic acid-modulated chondrocytes to reexpress the differentiated collagen phenotype without a change in cell shape. *Journal of Cell Biology* **106**, 161–70.

Borgens, R. B., Vanable, J. W. & Jaffe, L. F. (1977). Bioelectricity and regeneration II: large currents leave the stumps of regenerating newt limbs. *Proceedings of the National Academy of Sciences USA* **74**, 4528–32.

Brighton, C. T. Cronkey, J. E. & Osterman, A. L. (1976). *In vitro* epiphyseal plate growth in various constant electrical fields. *Journal of Bone and Joint Surgery* **58-A**, 971–8.

Brighton, C. T. & McCluskey, W. P. (1986). Cellular response and mechanisms of action of electrically induced osteogenesis. *Bone and Mineral Research* **4**, 213–54.

Brighton, C. T., Unger, A. & Stambough, J. (1984). *In vitro* growth of bovine articular cartilage in various capacitatively coupled electrical fields. *Journal of Orthopaedic Research* **2**, 15–22.

Clark, L. & Seawright, A. A. (1968). Skeletal abnormalities in the hind limbs of young cats as a result of hypervitaminosis A. *Nature* **217**, 1174–6.

Colacicco, G. & Pilla, A. (1984). Electromagnetic modulation of biological processes: influence of culture media and significance of methodology on the calcium uptake by embryonal chick tibia *in vitro*. *Calcified Tissue International* **36**, 167–74.

Conrad, G. W. & Dorfman, A. (1974). Synthesis of sulphated mucopolysaccharides by chick corneal fibroblasts *in vitro*. *Experimental Eye Research* **18**, 421–33.

Conrad, G. W., Hamilton, C. & Haynes, E. (1977). Differences in

glycosaminoglycans synthesized by fibroblast-like cells from chick cornea, heart, and skin. *Journal of Biological Chemistry* **252**, 6861–70.
Davidovitch, Z., Shanfeld, J. L., Montgomery, P. C., Lally, E., Laster, L., Furst, L. & Korosoff, E. (1984). Biochemical mediators of the effects of mechanical forces and electrical currents on mineralized tissues. *Calcified Tissue International* **36**, S86–97.
de la Haba, G. & Holtzer H. (1965). Chondroitin sulfate: inhibition of synthesis by puromycin. *Science* **149**, 1263–5.
Endo, N., Takahashi, H., Hiraki, Y., Takigawa, M. & Suzuki, F. (1984). Sensitization of rabbit costal chondrocytes to PTH by pulsing electromagnetic field stimulation. *Transactions of the Bioelectrical Repair and Growth Society* **4**, 8.
Farndale, R. (1982). Stimulation of matrix production by pulsed magnetic fields. *Transactions of the Bioelectrical Repair and Growth Society* **2**, 45.
Fitton-Jackson, S. F. & Bassett, C. A. L. (1981). The response of skeletal tissues to pulsed magnetic fields. In *Tissue Culture in Medical Research II*, ed. R. J. Richards & K. T. Rajan, pp. 21–2. Oxford: Pergamon Press.
Fitton-Jackson, S., Jones, D., Murray, J. & Farndale, R. (1981). The response of connective and skeletal tissue to pulsed magnetic fields. *Transactions of the Bioelectrical Repair and Growth Society* **1**, 85.
Fitzsimmons, R., Farley, J., Adey, W. R. & Baylink, D. (1986). Embryonic bone matrix formation is increased after exposure to a low amplitude capacitatively coupled electrical field *in vitro*. *Biochimica et Biophysica Acta* **882**, 51–6.
Friedenberg, Z. B., Harlow, M. C. & Brighton, C. T. (1971). Healing of nonunion of the medial malleolus by means of direct current. A case report. *Journal of Trauma* **11**, 883–5.
Hardingham, T. & Bayliss, M. (1990). Proteoglycans of articular cartilage: changes in aging and joint disease. *Seminars in Arthritis and Rheumatism* **20**, 12–33.
Heinegård, D. & Oldberg, A. (1989). Structure and biology of cartilage and bone noncollagenous macromolecules. *FASEB Journal* **3**, 2042–51.
Hiraki, Y., Endo, N., Takigawa, M., Asada, A., Takahashi, H. & Suzuki, F. (1987). Enhanced responsiveness to parathyroid hormone and induction of functional differentiation of cultured rabbit costal chondrocytes by a pulsed electromagnetic field. *Biochimica et Biophysica Acta* **931**, 94–100.
Holtzer, H., Abbott, J., Lash, J. & Holtzer, S. (1960). The loss of phenotypic traits by differentiated cells *in vitro*. I. Dedifferentiation of cartilage cells. *Proceedings of the National Academy of Sciences USA* **46**, 1533–42.
Horton, W. & Hassell, J. R. (1986). Independence of cell shape and

loss of cartilage matrix production during retinoic acid treatment of cultured chondrocytes. *Developmental Biology* **115**, 392–7.

Horton, W. E., Yamada, Y. & Hassell, J. R. (1987). Retinoic acid rapidly reduces articular cartilage matrix synthesis by altering gene transcription in chondrocytes. *Developmental Biology* **123**, 508–16.

Jaffe, L. F. (1979). The control of development by ionic currents. *Society for General Physiology Symposium* **33**, 199–231.

Johnson, D. & Rodan, G. (1982). The effect of pulsating electromagnetic fields on prostaglandin synthesis in osteoblast-like cells. *Transactions of the Bioelectrical Repair and Growth Society* **2**, 7.

Korenstein, R., Somjen, D., Danon, A., Fischler, H. & Binderman, I. (1981). Pulsed capacitative electric induction of cyclic-AMP changes, calcium^{45} uptake and DNA synthesis in bone cells. *Transactions of the Bioelectrical Repair and Growth Society* **1**, 34.

Laub, F. & Korenstein, R. (1984). Actin polymerization induced by pulsed electric stimulation of bone cells *in vitro*. *Biochimica et Biophysica Acta* **803**, 308–13.

McDevitt, C. A. & Muir, H. (1976). Biochemical changes in the cartilage of the knee in experimental and natural osteoarthrosis in the dog. *Journal of Bone and Joint Surgery* **58-B**, 94–101.

Norton, L. A. (1982). Effects of a pulsed electromagnetic field on a mixed chondroblastic tissue culture. *Clinical Orthopaedics and Related Research* **167**, 280–90.

Norton, L. A. (1985). Pulsed electromagnetic field effects on chondroblast culture. *Reconstructive Surgery and Traumatology* **19**, 70–86.

Norton, L. A., Rodan, G. A. & Bourret, L. A. (1977). Epiphyseal cartilage cAMP changes produced by electrical and mechanical perturbations. *Clinical Orthopaedics* **124**, 59–68.

Norton, L. A. & Rovetti, L. A. (1988). Calcium incorporation in cultured chondroblasts perturbed by an electromagnetic field. *Journal of Orthopaedic Research* **6**, 559–66.

Pelletier, J. P., Martel-Pelletier, J., Mehraban, F. & Malemud, C. J. (1992). Immunological analysis of proteoglycan structural changes in the early stage of experimental osteoarthritic canine cartilage lesions. *Journal of Orthopaedic Research* **10**, 511–23.

Sakai, A., Suzuki, K., Nakamura, T., Norimura, T. & Tsuchiya, T. (1991). Effects of pulsing electromagnetic fields on cultured cartilage cells. *International Orthopaedics* **15**, 341–6.

Schiller, S., Slover, G. A. & Dorfman, A. (1961). A new method for the separation of acid mucopolysaccharides: its application to the isolation of heparin from the skin of rats. *Journal of Biological Chemistry* **236**, 983–7.

Shamos, M. H., Lavine, L. S. & Shamos, M. I. (1963). Piezoelectric effect in bone. *Nature* **197**, 81.

Smith, R. (1984). Reduction of articular destruction in infectious arthritis by pulsing electromagnetic fields. *Transactions of the Bioelectrical Repair and Growth Society* **4**, 41.

Smith, R. L. & Nagel, D. A. (1983). Effects of pulsing electromagnetic fields on bone growth and articular cartilage. *Clinical Orthopaedics and Related Research* **181**, 277–82.

Yasuda, I. (1953). Fundamental aspects of fracture treatment. *Journal of the Kyoto Medical Society* **4**, 395–406.

Index